Andreas Graef

**Statistical Analysis of Brain Signals for Epileptology**

Andreas Graef

# Statistical Analysis of Brain Signals for Epileptology

## Automated Procedures for Focus Detection and Seizure Propagation Analysis from Ictal Electrocorticography

**Südwestdeutscher Verlag für Hochschulschriften**

**Impressum / Imprint**

Bibliografische Information der Deutschen Nationalbibliothek: Die Deutsche Nationalbibliothek verzeichnet diese Publikation in der Deutschen Nationalbibliografie; detaillierte bibliografische Daten sind im Internet über http://dnb.d-nb.de abrufbar.

Alle in diesem Buch genannten Marken und Produktnamen unterliegen warenzeichen-, marken- oder patentrechtlichem Schutz bzw. sind Warenzeichen oder eingetragene Warenzeichen der jeweiligen Inhaber. Die Wiedergabe von Marken, Produktnamen, Gebrauchsnamen, Handelsnamen, Warenbezeichnungen u.s.w. in diesem Werk berechtigt auch ohne besondere Kennzeichnung nicht zu der Annahme, dass solche Namen im Sinne der Warenzeichen- und Markenschutzgesetzgebung als frei zu betrachten wären und daher von jedermann benutzt werden dürften.

Bibliographic information published by the Deutsche Nationalbibliothek: The Deutsche Nationalbibliothek lists this publication in the Deutsche Nationalbibliografie; detailed bibliographic data are available in the Internet at http://dnb.d-nb.de.

Any brand names and product names mentioned in this book are subject to trademark, brand or patent protection and are trademarks or registered trademarks of their respective holders. The use of brand names, product names, common names, trade names, product descriptions etc. even without a particular marking in this works is in no way to be construed to mean that such names may be regarded as unrestricted in respect of trademark and brand protection legislation and could thus be used by anyone.

Coverbild / Cover image: www.ingimage.com

Verlag / Publisher:
Südwestdeutscher Verlag für Hochschulschriften
ist ein Imprint der / is a trademark of
OmniScriptum GmbH & Co. KG
Heinrich-Böcking-Str. 6-8, 66121 Saarbrücken, Deutschland / Germany
Email: info@svh-verlag.de

Herstellung: siehe letzte Seite /
Printed at: see last page
**ISBN: 978-3-8381-3788-9**

Zugl. / Approved by: Wien, TU, Diss., 2013

Copyright © 2014 OmniScriptum GmbH & Co. KG
Alle Rechte vorbehalten. / All rights reserved. Saarbrücken 2014

# Preface

This book represents a slightly revised version of my doctoral thesis submitted at Vienna University of Technology (Graef 2013), with some minor adaptations and corrections of evident mistakes.

I wish to thank all researchers who contributed to this study, in particular Prof. Manfred Deistler from Vienna University of Technology and Prof. Christoph Baumgartner from Hospital Hietzing with Neurological Center Rosenhügel, Vienna. Special thanks go to my colleague Dr. Christoph Flamm, to Dr. Susanne Pirker, head of the Epilepsy Monitioring Unit at Neurological Center Rosenhügel, to Prof. Gerald Matz from Vienna University of Technology, Institute of Telecommunications, and to Dr. Tilmann Kluge and Manfred Hartmann from AIT Austrian Institute of Technology.

I am thankful to Anke Metzger from Südwestdeutscher Verlag für Hochschulschriften for the smooth collaboration leading to this publication.

This study has been approved by the ethics committee of the City of Vienna (vote EK 12-100-0712). Signed informed patient consents permitting use of anonymized data for research and publication are available.
It was supported by the Austrian Science Fund FWF (grant P22961).

*Andreas Graef*
Vienna, February 2014

Затем вдруг как бы что-то разверзлось пред ним: необычайный внутренний свет озарил его душу. Это мгновение продолжалось, может быть, полсекунды; но он, однако же, ясно и сознательно помнил начало, самый первый звук своего страшного вопля, который вырвался из груди его сам собой и который никакою силой он не мог бы остановить. Затем сознание его угасло мгновенно, и наступил полный мрак. С ним случился припадок эпилепсии, уже очень давно оставившей его.

— Ф. М. Достоевский: Идиот

Next moment something appeared to burst open before him: a wonderful inner light illuminated his soul. This lasted perhaps half a second, yet he distinctly remembered hearing the beginning of the wail, the strange, dreadful wail, which burst from his lips of its own accord, and which no effort of will on his part could suppress. Next moment he was absolutely unconscious; black darkness blotted out everything. He had fallen in an epileptic fit.

— *The Idiot*[1] by F. M. Dostoyevsky (1821-1881), one of the most famous epilepsy patients.

---

[1] Part II, chapter V; English translation by Eva Martin

# Contents

| Contents | 3 |
| --- | --- |
| List of Figures | 11 |
| List of Tables | 13 |

## I  Introduction   15

**1  Introduction**   17
   1.1  Motivation . . . . . . . . . . . . . . . . . . . . . . . .  17
   1.2  History of epilepsy  . . . . . . . . . . . . . . . . . . .  18
        1.2.1  Historical overview . . . . . . . . . . . . . . . .  18
            1.2.1.1  Ancient history . . . . . . . . . . . .  18
            1.2.1.2  Medieval history . . . . . . . . . . .  20
            1.2.1.3  Modern history . . . . . . . . . . . .  20
        1.2.2  Famous epilepsy patients . . . . . . . . . . . . .  23
   1.3  Outline . . . . . . . . . . . . . . . . . . . . . . . . . .  24
   1.4  Notation . . . . . . . . . . . . . . . . . . . . . . . . .  25

**2  Medical Background**   27
   2.1  Human brain . . . . . . . . . . . . . . . . . . . . . . .  27
        2.1.1  Anatomy  . . . . . . . . . . . . . . . . . . . . .  27
        2.1.2  Physiology . . . . . . . . . . . . . . . . . . . . .  30
   2.2  Electroencephalography . . . . . . . . . . . . . . . . .  32
        2.2.1  Definition . . . . . . . . . . . . . . . . . . . . .  33
        2.2.2  Recording conventions . . . . . . . . . . . . . .  34
        2.2.3  Localization of discharges . . . . . . . . . . . .  36

|         |       |         |                                          |     |
|---------|-------|---------|------------------------------------------|-----|
|         | 2.2.4 | Artifacts ............................... | 39 |
|         | 2.2.5 | Invasive EEG ............................ | 42 |
| 2.3 | Epilepsies ........................................ | 44 |
|         | 2.3.1 | Definitions ............................. | 44 |
|         | 2.3.2 | Classification .......................... | 46 |
|         | 2.3.3 | Diagnosis ............................... | 49 |
|         | 2.3.4 | Therapy ................................. | 51 |
|         | 2.3.5 | Localizing value of EEG ................ | 52 |
|         |       | 2.3.5.1 Concept of the cortical zones .... | 53 |
|         |       | 2.3.5.2 Ictal rhythmic patterns .......... | 54 |
|         |       | 2.3.5.3 Interictal spikes ................ | 54 |
|         |       | 2.3.5.4 High frequency oscillations ...... | 55 |
|         |       | 2.3.5.5 Ictal slow shifts ................ | 56 |
|         | 2.3.6 | Epilepsy surgery ........................ | 57 |
|         |       | 2.3.6.1 Presurgical evaluation ........... | 57 |
|         |       | 2.3.6.2 Surgical intervention ............ | 59 |
|         |       | 2.3.6.3 Postsurgical outcome ............. | 60 |
|         | 2.3.7 | Social aspects .......................... | 62 |

# 3 Statistical Background     63

| 3.1 | Non-parametric spectral estimation ............. | 63 |
|-----|-------------------------------------------------|----|
|     | 3.1.1 Typical issues ........................... | 65 |
|     | 3.1.2 Preliminary definitions ................. | 67 |
|     | 3.1.3 Indirect estimation methods ............. | 70 |
|     | 3.1.4 Direct estimation methods ............... | 71 |
|     |       3.1.4.1 Periodogram ................... | 71 |
|     |       3.1.4.2 Welch power spectral estimation . | 72 |
| 3.2 | Autoregressive modeling ........................ | 74 |
| 3.3 | Granger causality .............................. | 76 |
|     | 3.3.1 Conditional Granger Causality ........... | 77 |
|     | 3.3.2 Granger Causality in the frequency domain | 78 |
| 3.4 | Dependency measures ............................ | 78 |

|   |   | 3.4.1 | Ordinary Coherence . . . . . . . . . . . . . . . . . | 80 |
|---|---|---|---|---|
|   |   | 3.4.2 | Partial Coherence . . . . . . . . . . . . . . . . . . | 81 |
|   |   | 3.4.3 | Directed Transfer Function . . . . . . . . . . . . | 83 |
|   |   |   | 3.4.3.1 DTF . . . . . . . . . . . . . . . . . . . | 83 |
|   |   |   | 3.4.3.2 Extensions . . . . . . . . . . . . . . . | 85 |
|   |   |   | 3.4.3.3 Discussion . . . . . . . . . . . . . . . | 86 |
|   |   | 3.4.4 | Partial Directed Coherence . . . . . . . . . . . . | 87 |
|   |   |   | 3.4.4.1 PDC . . . . . . . . . . . . . . . . . . . | 87 |
|   |   |   | 3.4.4.2 Extensions . . . . . . . . . . . . . . . | 91 |
|   |   |   | 3.4.4.3 Discussion . . . . . . . . . . . . . . . | 91 |
|   |   | 3.4.5 | Granger Causality Index . . . . . . . . . . . . . | 92 |
|   | 3.5 | Signal detection . . . . . . . . . . . . . . . . . . . . . . . . | | 93 |
|   |   | 3.5.1 | Preliminary definitions . . . . . . . . . . . . . . | 94 |
|   |   | 3.5.2 | Simple hypothesis testing . . . . . . . . . . . . . | 96 |
|   |   | 3.5.3 | Composite hypothesis testing . . . . . . . . . . . | 97 |
|   |   |   | 3.5.3.1 Uniformly Most Powerful Tests . . . . . | 97 |
|   |   |   | 3.5.3.2 Invariance . . . . . . . . . . . . . . . | 98 |
|   |   |   | 3.5.3.3 Matched subspace filters . . . . . . . . | 101 |
|   | 3.6 | Factor Models . . . . . . . . . . . . . . . . . . . . . . . . . | | 105 |

## II Materials and Methods 109

## 4 Propagation Analysis Framework 111

|   | 4.1 | Introduction . . . . . . . . . . . . . . . . . . . . . . . . . . | | 111 |
|---|---|---|---|---|
|   |   | 4.1.1 | Medical background . . . . . . . . . . . . . . . | 112 |
|   |   | 4.1.2 | Technical background . . . . . . . . . . . . . . . | 113 |
|   | 4.2 | Framework . . . . . . . . . . . . . . . . . . . . . . . . . . | | 115 |
|   | 4.3 | Data . . . . . . . . . . . . . . . . . . . . . . . . . . . . . . | | 118 |
|   |   | 4.3.1 | Patient history . . . . . . . . . . . . . . . . . . | 118 |
|   |   | 4.3.2 | ECoG recordings . . . . . . . . . . . . . . . . . | 118 |
|   |   | 4.3.3 | Visual inspection . . . . . . . . . . . . . . . . . | 121 |

# 5 HFO Detection        123

  5.1 Introduction . . . . . . . . . . . . . . . . . . . . . . . . 123
      5.1.1 Background . . . . . . . . . . . . . . . . . . . . 123
      5.1.2 Contribution . . . . . . . . . . . . . . . . . . . . 125
  5.2 Method . . . . . . . . . . . . . . . . . . . . . . . . . . . 126
      5.2.1 Matched subspace filter . . . . . . . . . . . . . . 126
      5.2.2 Signal model . . . . . . . . . . . . . . . . . . . 130
  5.3 Results . . . . . . . . . . . . . . . . . . . . . . . . . . . 130
      5.3.1 Signal model . . . . . . . . . . . . . . . . . . . 130
      5.3.2 HFO propagation . . . . . . . . . . . . . . . . . 132
  5.4 Discussion . . . . . . . . . . . . . . . . . . . . . . . . . 134
      5.4.1 Matched subspace filter . . . . . . . . . . . . . . 134
      5.4.2 HFO propagation . . . . . . . . . . . . . . . . . 136
      5.4.3 Concluding remarks . . . . . . . . . . . . . . . . 138

# 6 Causality Analysis        139

  6.1 Introduction . . . . . . . . . . . . . . . . . . . . . . . . 139
      6.1.1 Background . . . . . . . . . . . . . . . . . . . . 139
      6.1.2 Contribution . . . . . . . . . . . . . . . . . . . . 140
  6.2 Method . . . . . . . . . . . . . . . . . . . . . . . . . . . 141
      6.2.1 Autoregressive model . . . . . . . . . . . . . . . 141
      6.2.2 Solution of the normal equations . . . . . . . . . 142
      6.2.3 Dynamic input channel selection . . . . . . . . . 143
      6.2.4 Partial extrinsic power . . . . . . . . . . . . . . 145
      6.2.5 Extrinsic-to-intrinsic-power-ratio (EIPR) . . . . . 146
      6.2.6 Comparison of EIPR and PDC . . . . . . . . . . 147
      6.2.7 Signal model . . . . . . . . . . . . . . . . . . . 149
  6.3 Results . . . . . . . . . . . . . . . . . . . . . . . . . . . 151
      6.3.1 Signal model . . . . . . . . . . . . . . . . . . . 151
      6.3.2 Analysis of the channel selection algorithm . . . 152
      6.3.3 Seizure onset zone localization . . . . . . . . . . 153
  6.4 Discussion . . . . . . . . . . . . . . . . . . . . . . . . . 156

|  |  | 6.4.1 | EIPR as coupling indicator . . . . . . . . . . . . . 156 |
|  |  | 6.4.2 | Behavior of the channel selection algorithm . . . . 161 |
|  |  | 6.4.3 | Seizure onset zone localization . . . . . . . . . . . 162 |
|  |  | 6.4.4 | Concluding remarks . . . . . . . . . . . . . . . . . 164 |

**7 Influence Analysis**     **167**
- 7.1 Introduction . . . . . . . . . . . . . . . . . . . . . . . . . . 167
  - 7.1.1 Background . . . . . . . . . . . . . . . . . . . . . 167
  - 7.1.2 Granger causality for factor models . . . . . . . . 168
- 7.2 Method . . . . . . . . . . . . . . . . . . . . . . . . . . . . 170
  - 7.2.1 Proposed methodology . . . . . . . . . . . . . . . 170
  - 7.2.2 Signal model . . . . . . . . . . . . . . . . . . . . . 173
- 7.3 Results . . . . . . . . . . . . . . . . . . . . . . . . . . . . 174
  - 7.3.1 Signal model . . . . . . . . . . . . . . . . . . . . . 174
  - 7.3.2 Seizure onset zone localization . . . . . . . . . . . 178
- 7.4 Discussion . . . . . . . . . . . . . . . . . . . . . . . . . . 180
  - 7.4.1 Influence analysis . . . . . . . . . . . . . . . . . . 180
  - 7.4.2 Seizure onset zone localization . . . . . . . . . . . 184
  - 7.4.3 Concluding remarks . . . . . . . . . . . . . . . . . 186

**8 Segmentation**     **189**
- 8.1 Introduction . . . . . . . . . . . . . . . . . . . . . . . . . 189
  - 8.1.1 Background . . . . . . . . . . . . . . . . . . . . . 189
  - 8.1.2 Contribution . . . . . . . . . . . . . . . . . . . . . 192
- 8.2 Method . . . . . . . . . . . . . . . . . . . . . . . . . . . . 193
  - 8.2.1 Segmentation . . . . . . . . . . . . . . . . . . . . 193
  - 8.2.2 Periodic waveform analysis . . . . . . . . . . . . . 194
  - 8.2.3 Segment classification . . . . . . . . . . . . . . . 196
  - 8.2.4 SOZ localization and seizure propagation . . . . . 198
  - 8.2.5 Signal model . . . . . . . . . . . . . . . . . . . . . 198
- 8.3 Results . . . . . . . . . . . . . . . . . . . . . . . . . . . . 200
  - 8.3.1 Segmentation of simulated data . . . . . . . . . . 201
  - 8.3.2 Segmentation of ECoG data . . . . . . . . . . . . 202

|  |  | 8.3.3 | Classification of ECoG segments . . . . . . . . . . 203 |
|  |  | 8.3.4 | Onset zone analysis . . . . . . . . . . . . . . . . 204 |
|  | 8.4 | Discussion . . . . . . . . . . . . . . . . . . . . . . . . . . . . 206 |
|  |  | 8.4.1 | Segmentation . . . . . . . . . . . . . . . . . . . 206 |
|  |  | 8.4.2 | Classification . . . . . . . . . . . . . . . . . . . 210 |
|  |  | 8.4.3 | Seizure propagation . . . . . . . . . . . . . . . . 210 |
|  |  | 8.4.4 | Concluding remarks . . . . . . . . . . . . . . . . 211 |

# III Results and Discussion 213

## 9 Framework results 215
9.1 Introduction . . . . . . . . . . . . . . . . . . . . . . . . . . . 215
9.2 Previous clinical findings . . . . . . . . . . . . . . . . . . . 216
9.3 Analysis of seizures . . . . . . . . . . . . . . . . . . . . . . 218
9.4 Overall findings . . . . . . . . . . . . . . . . . . . . . . . . 220

## 10 Discussion and Outlook 223
10.1 Patient . . . . . . . . . . . . . . . . . . . . . . . . . . . . . . 223
10.2 Framework . . . . . . . . . . . . . . . . . . . . . . . . . . . 227
    10.2.1 Limitation to ECoG data . . . . . . . . . . . . . 227
    10.2.2 Performance . . . . . . . . . . . . . . . . . . . . 232
10.3 Outlook . . . . . . . . . . . . . . . . . . . . . . . . . . . . . 234

# IV Appendix 237

## A Proofs for the statistical background 239
A.1 Non-parametric spectral estimation . . . . . . . . . . . . . 239
    A.1.1 Preliminary definitions . . . . . . . . . . . . . . . 239
    A.1.2 Indirect estimation methods . . . . . . . . . . . . 240
    A.1.3 Direct estimation methods . . . . . . . . . . . . . 241
        A.1.3.1 Periodogram . . . . . . . . . . . . . . . 241
        A.1.3.2 Welch method . . . . . . . . . . . . . . 245

|  |  |  |
|---|---|---|
| A.2 | Granger causality | 250 |
| A.3 | Dependency measures | 250 |
|  | A.3.1 Partial Coherence | 250 |
|  | A.3.2 Directed Transfer Function | 252 |
|  | A.3.3 Partial Directed Coherence | 253 |
| A.4 | Signal detection | 255 |

**Bibliography**     259

**Index**     303

# List of Figures

| | | |
|---|---|---|
| 2.1.1 | Schematic representation of the cerebrum . . . . . . . . | 28 |
| 2.1.2 | Neural information transmission . . . . . . . . . . . . | 31 |
| 2.2.1 | EEG recording conventions . . . . . . . . . . . . . . . | 35 |
| 2.2.2 | Exemplary EEG recordings . . . . . . . . . . . . . . . | 37 |
| 2.2.3 | Typical physiological artifacts . . . . . . . . . . . . | 40 |
| 2.2.4 | Typical technical artifacts . . . . . . . . . . . . . . | 41 |
| 3.1.1 | Commonly used windows . . . . . . . . . . . . . . . . . | 69 |
| 3.4.1 | DTF and Granger causality. . . . . . . . . . . . . . . | 86 |
| 3.4.2 | Normalization of the PDC. . . . . . . . . . . . . . . . | 89 |
| 3.5.1 | Receiver Operator Characteristics . . . . . . . . . . . | 96 |
| 3.5.2 | Symmetry of linear data model . . . . . . . . . . . . . | 102 |
| 3.5.3 | Matched subspace filter . . . . . . . . . . . . . . . . | 104 |
| 4.3.1 | MRI scan with electrode positions . . . . . . . . . . . | 119 |
| 4.3.2 | Seizure 1 . . . . . . . . . . . . . . . . . . . . . . . | 120 |
| 4.3.3 | Seizure 2 . . . . . . . . . . . . . . . . . . . . . . . | 120 |
| 4.3.4 | Seizure 3 . . . . . . . . . . . . . . . . . . . . . . . | 121 |
| 5.3.1 | HFO detection: Simulation results . . . . . . . . . . . | 131 |
| 5.3.2 | HFO detection: results for seizure 1 . . . . . . . . . | 131 |
| 5.3.3 | HFO propagation . . . . . . . . . . . . . . . . . . . . | 133 |
| 5.4.1 | HFO detection via RMS in seizure 1 . . . . . . . . . . | 136 |
| 5.4.2 | HFO propagation in seizure 1 . . . . . . . . . . . . . | 137 |
| 5.4.3 | SOZ according to HFO detection . . . . . . . . . . . . | 137 |
| 6.2.1 | Automatic channel selection algorithm . . . . . . . . . | 144 |
| 6.2.2 | Dependence graphs of AR models . . . . . . . . . . . . | 149 |

| | | |
|---|---|---|
| 6.3.1 | Arrow maps based on EIPR | 157 |
| 6.3.2 | Out-degrees of arrow maps based on EIPR | 158 |
| 6.4.1 | PDC matrix plot of the signal model | 159 |
| 6.4.2 | Zoom into two subplots of the PDC matrix-plot | 161 |
| 6.4.3 | SOZ according to causality analysis | 163 |
| 7.2.1 | Visualization of the influence analysis | 172 |
| 7.2.2 | Illustration of the dependence structure of signal model | 175 |
| 7.3.1 | Scree plot | 175 |
| 7.3.2 | Results of the influence analysis of signal model | 176 |
| 7.3.3 | Results of the influence analysis of seizure 1 | 177 |
| 7.3.4 | Arrow maps based on influence analysis | 181 |
| 7.3.5 | Out-degrees of arrow maps based on influence analysis | 182 |
| 7.4.1 | SOZ according to influence analysis | 185 |
| 8.2.1 | PWI and coupled frequency | 196 |
| 8.3.1 | Segmentation of simulated data | 201 |
| 8.3.2 | Segmentation of channel A12, seizure 1 | 203 |
| 8.3.3 | Segment classification of channel A12, seizure 1 | 204 |
| 8.3.4 | SOZ and initial spread of epileptic activity | 205 |
| 8.3.5 | SOZ according to segmentation method | 206 |
| 8.4.1 | Segmentation behavior preictally and ictally | 209 |
| 8.4.2 | Influence of threshold on segment length | 209 |
| 9.3.1 | SOZ according to seizure analysis | 220 |
| 9.4.1 | Seizure propagation according to framework | 221 |
| 10.1.1 | Depth electrode positions | 225 |
| 10.1.2 | Intracerebral EEG | 226 |
| 10.2.1 | EOG artifact correction | 230 |
| 10.2.2 | Low spatial resolution of surface EEG | 231 |

# List of Tables

| | | |
|---|---|---:|
| 2.2.1 | Localization rules in EEG | 39 |
| 2.3.1 | ILAE classification 2009 of seizures | 47 |
| 2.3.2 | ILAE classification 2009 of epilepsies | 48 |
| 2.3.3 | Classification of post-surgical outcome | 60 |
| 3.1.1 | Characteristics of commonly used windows | 69 |
| 4.2.1 | Seizure propagation analysis framework | 117 |
| 4.3.1 | Visual inspection of the ECoG raw data | 122 |
| 5.1.1 | Overview of basic HFO detection approaches | 124 |
| 5.3.1 | HFO propagation | 134 |
| 6.3.1 | Values of EIPR for signal model (6.2.11) | 152 |
| 6.3.2 | Step-wise behavior of the channel selection algorithm | 154 |
| 6.3.3 | SOZ according to causality analysis | 155 |
| 7.3.1 | SOZ according to influence analysis | 179 |
| 8.1.1 | Overview of segmentation strategies | 191 |
| 8.2.1 | Segment classification | 199 |
| 8.3.1 | SOZ according to segmentation method | 207 |
| 9.3.1 | Detailed analysis of the three seizures | 219 |
| 9.3.2 | Seizure propagation according to seizure analysis | 219 |
| 10.1.1 | Seizure onset in intracerebral recordings | 227 |
| 10.2.1 | Recent automatic artifact removal approaches | 229 |

# Part I

# Introduction

# 1 Introduction

## 1.1 Motivation

> Advancing medicine and biology through the application of engineering sciences and technology
>
> — Part of the mission statement of the *IEEE EMBS (Engineering in Medicine and Biology Society)*[1]

In this study we are concerned with the automated analysis of epileptic seizures, the manifestations of one of the most common neurological diseases, epilepsy. For this purpose we combine technical methods with classical clinical approaches. This field of research is often referred to as *Biomedical Engineering*. Its aim is the use of engineering technology in medicine for the sake of patient's cure.

The goal of this study is to develop novel technological methods for epileptic seizure propagation analysis which have the potential for future clinical use. Therefore, this work is guided by theoretical considerations, but the emphasis is strictly put on application. In the author's view, the derivation of theoretical results and the subsequent validation by simulations is only a prerequisite for the final step: the test on patient data.

Formal elegance of a technological method is nice to have, but what makes it »beautiful« in the end is clinical usability. We will see that surprisingly the mathematically simplest of our methods perform best.

In this study we successfully test our methodology on data of one epilepsy patient. Thus, the material presented in here has to be regarded as a first step towards clinical usability. A roll-out on a larger group of patients,

---

[1] See the »About EMBS« section on the homepage of the IEEE EMBS, www.embs.org/about-embs/what-we-do.

e.g. at least five, is the next necessary step, but not within the scope of this study. Neither are in-depth considerations of mathematical theory behind our framework; we refer to the doctoral thesis of Flamm (2012) for this purpose.

## 1.2 History of epilepsy

> Περὶ μὲν τῆς ἱερῆς νούσου καλεομένης ὧδ᾽ ἔχει: οὐδέν τί μοι δοκέει τῶν ἄλλων θειοτέρη εἶναι νούσων οὐδὲ ἱερωτέρη, ἀλλὰ φύσιν μὲν ἔχει ἣν καὶ τὰ λοιπὰ νουσήματα, ὅθεν γίνεται.
>
> — Ἱπποκράτης: *Περὶ ἱερῆς νούσου*
>
> It is thus with regard to the disease called Sacred: it appears to me to be nowise more divine nor more sacred than other diseases, but has a natural cause from the originates like other affections.
>
> — Hippocrates: *On the Sacred Disease*, first epileptologic monograph (ca. 400 BC)[2]

In this section we give a short overview of the historical developments in social reception, diagnosis and therapy of epilepsies as well as a list of famous epilepsy patients. This whole section is based on Schneble (2003).

### 1.2.1 Historical overview

Hardly any other disease can be traced back so far in history like epilepsy. The reason for this transparency most probably lies in its high prevalence and the dramatic clinical symptomatology.

#### 1.2.1.1 Ancient history

Oldest written evidence of epilepsy dates back to the 17th century BC. Ancient Egyptian hieroglyphics (beginning of the 17th dynasty, 1650-1570

---

[2]Opening of Section 1; English translation by Francis Adams.
 Original text and English translation are accessible online via the PERSEUS project at www.perseus.tufts.edu (»de morbo sacro«).

BC) mention a disease termed »nesejet« which was believed to be divine. The Codex Hammurabi (17th century BC), §278, assures a warranty clause of 100 days for slaves suffering from the »benu« disease. Finally, a plate of Babylonian cuneiform writing (ca. 1050 BC), which is part of a series of plates serving as an Ancient Babylonian medical textbook, forms an integral chapter on epilepsy.

The Ancient Chinese medicine knows about epilepsy as well, providing descriptions (Tschou dynasty, 770-221 BC) and classification attempts (Dui dynasty, 610 AC).

The Old Testament establishes a link between epilepsy and prophecy, compare Feininger (2000). Although a strict technical term for epilepsy is missing, it provides several evidences including David, Saul and the prophet Bileam who refers to himself as »falling« (»nôphél«; Numeri, 24, 1-14; 15-16).

The Talmud speaks about epilepsy more often, but mostly with regard to the social context of the patient.

The Ancient Greek medicine led to a revolution in epileptology. Hippocrates (ca. 460-377 BC) published the first epileptologic monograph »On the Sacred Disease« in which he argues that epilepsy is not divine, but has natural causes (compare the opening citation of this section). He elaborates this argument by explaining that »men regard its nature and cause as divine from ignorance and wonder, because it is not at all like to other diseases. And this notion of its divinity is kept up by their inability to comprehend it, and the simplicity of the mode by which it is cured, for men are freed from it by purifications and incantations.«[3] In this text Hippocrates establishes a connection between the Ancient Greek word »epilambanein« (ἐπιλαμβάνειν = to seize/attack) and the manifestation of the disease, therefrom the modern term epilepsy.

For the first time in history, Hippocrates considered malfunctions of the

---

[3] Section 1; English translation by Francis Adams.
Original text and English translation are accessible online via the PERSEUS project at www.perseus.tufts.edu (»de morbo sacro«).

»physis« of the brain, i.e. what characterizes the brain as origin of rational thinking, as the reason for epilepsy. Although this approach seems to be very near to modern medicine, the ancient medical concept of humorism developed by him is not: Hippocrates believed that certain human moods, emotions and behaviors were caused by body fluids (called »humors«, from the Greek word χυμός): blood, yellow bile, black bile, and phlegm. According to this school of thought, seizures are provoked if humors mix in a wrong way, thus disturbing the equilibrium state in the brain.

As Ancient Roman medicine was heavily built on the Greek one, the Greek body of thought was incorporated into the Roman epileptology. Aulus Cornelius Celsus published the first medical text about epilepsy in Latin (»De medicina libri octo«, ca. 30 BC), using the terms »morbus maior« and »morbus comitialis«. Claudius Galen (ca. 129-201 AC) extended Hippocrates' humorism by mapping the four temperaments (temperare = to mix) to the adjectives hot/cold/dry/wet. Based on this theory, he provided a classification and description of epilepsy.

### 1.2.1.2 Medieval history

As in many other fields of science, the middle ages led to a regression in epileptology, and the supernatural character of epilepsy became predominant again. Demons as the cause of epilepsy gave name to the term »morbus daemonicus«, curable by exorcism. Medieval society referred to the New Testament as proof for the non-natural character of epilepsy, where it is reported that Jesus healed a »moonstruck son« (Mt 17, 14ff; Lk 9, 38ff; Mk 9, 17ff). This argument gave rise to the term »morbus lunaticus«.

### 1.2.1.3 Modern history

Modern history starts with the period of Renaissance, characterized by a return to the ancient body of thought. Paracelsus' (Aureolus Philippus Theophrastus Bombastus von Hohenheim, 1493-1541) theory that actions in macrocosms have an analog in the microcosm, i.e. the human body, was applied to epilepsy as well. He explained the pathogenesis of epilepsy, using the term »morbus caducus«, by this *concordance principle*, e.g thunder corresponds to convulsions of an epilepsy patient. According to this theory, the three principles of the organism (sulfur, mercury and salt) influenced the pathonogenesis and the four elements (fire, air, earth, water) defined the severity of epileptic seizures.

It is interesting to note that throughout middle ages and still in Renaissance epilepsy was believed to be an infectious disease.

In the 17th century Francois de la Boe, also known as Sylvius von Leyden, claimed that the acidity ratio in the human body was an important indicator in medicine. In 1674 he published »de morbus infantum« in which he identified hyper-acid blood components as the cause of epilepsy. Consequently, he suggested a chemical therapy with alkali. Interestingly, the basis of this idea, namely the inequality of acids and alkali, still holds in modern medicine, but the opposite of his argument is true: The alkaline milieu due to e.g. hyperventilation can provoke seizures.

The 18th and beginning 19th centuries represent the start of modern epileptology. In 1770 Tissot published his »Traité de l'épilepsie« in which he classified epilepsies into idiopathic - symapathic - essential and introduced a symptomatology into »small« and »big« seizures: a scheme which should influence epileptology until the 20th century. In 1841, another famous epileptologist, father West, wrote a letter to the editor of Lancet describing the West syndrome in his son; an epilepsy syndrome still named after him.

The 19th century meant a break-through in epileptology: Epilepsy was finally recognized as cerebral disease, which led to the insight that epilepsy patients need clinical assistance and eventually care.

With natural sciences evolving quicker and quicker in the 19th century, the second half of this century represented a golden era for epileptology. Science understood the morphological and functional anatomy of the human brain and gained patho-physiological insights into the nervous system: In

1870 Gustav Theodor Fritsch and Eduard Hietzing managed to answer the old question whether seizures can only be generated cortically. In their study »Über die elektrische Erregbarkeit des Großhirns«, they demonstrated in animal experiments that the electrical stimulation of cortical areas triggers motor actions and that endured stimulation can provoke seizures. Only three years later, in 1873, John Hughlings-Jackson (who can be seen as the father of modern epileptology) gave a seminal definition of epilepsy in his study »On the anatomical, physiological, and pathological investigation of epilepsies«.

With the advances in chemistry, systematic epilepsy therapy started: After the discovery of bromine as element and its extraction from sea water in 1826 it became an all-round medication in the middle of the century. In the second half of the century the clinician Charles Locock presumed that it might also show an anti-epileptic effect – the medicinal therapy of epilepsy was born.

In the 20th century scientific progress still continued to speed up. The sensational progress in therapy and diagnostic of epilepsy led to the foundation of the *International League Against Epilepsy (ILAE)* in Budapest in 1909.

Medicinal therapy was boosted by the discovery of the anti-epileptic effect of phenobarbital in 1912, as up to that moment only bromine (with a large number of side effects) was known. Then, step by step, more anti-epileptic agents were synthesized in a systematic way: introduction of phenytoin in 1938, primidon in 1952, ethosuximid in 1958, valproic acid as omnipotent anti-epileptic agent in 1962, carbamazepine and other benzodiazepines in 1962. From that moment on waiting for new anti-epileptic agents should last until the 1990s. Research then brought a flood of new molecules (vigabatrin, lamotrigin, felbamat, gabapentin, tiagabin, topiramat, oxcarbazepine, levetiracetam; compare Subsection 2.3.4).

Diagnostic was revolutionized by the invention of the electroencephalogram (EEG), compare Section 2.2. Hans Berger managed to record brain waves intra-operatively in 1924 (first doubtless evidence), and his world-famous publication (Berger 1929) »Über das Elektrenkephalogramm des Menschen« followed immediately. Modern imaging techniques (CT, MRI, PET, SPECT, fMRI) then led to a second revolution in the whole field of medicine and are nowadays an inherent part of modern diagnostics in epileptology, compare Subsection 2.3.3.

Besides medicinal therapy the surgical treatment came up at the beginning of the 20th century, compare Subsection 2.3.6. Wilder Penfield (1891-1976), who worked in Montreal in the 1920/30s, performed the first epilepsy-surgical intervention: He removed a brain tumor of an adolescent suffering from focal epilepsy which successfully resulted in seizure remission (Penfield 1934). The cooperation with the neurophysiologist Herbert Jasper (1906-1999) increased the post-surgical outcome significantly: Jasper knew how to use Berger's EEG for presurgical focus localization, and together they developed the »Montreal method« based on intra-operative electrocorticographic (ECoG) monitoring.

### 1.2.2 Famous epilepsy patients

Epilepsy is one of the most common neurological diseases with a prevalence of 0.7% (Hirtz et al. 2007) in the general population. Therefore, many celebrities as well have been suffering from epilepsy throughout all epochs. Famous patients with a reliable diagnosis of epilepsy include (Schneble 2003)

**Persons of the Holy Bible:** Bileam (ca. 1500 BC), Saul (ca. 1000 BC) and the Holy Paul (ca. 10-67 AC)

**Emperors:** the Roman emperor Gaius Julius Caesar (100-44 BC), who suffered from epileptic seizures according to Sueton (»Comitiali quoque morbo bis inter res agendas correptus est.«); Napoleon Bonaparte (1769-1821) and his Austrian adversary Archduke Charles of Austria (1771-1847)

**Church representatives:** Cardinal Richelieu (1585-1642)

**Artists and writers:** Gustave Flaubert (1821-1880), Fjodor Michailovich Dostojevskij (1821-1881; epilepsy motives in two of his famous books, »The Idiot« and »Brothers Karamasov«; compare the introductory quote) and Vincent van Gogh (1853-1890)

Furthermore, indications for epilepsy can be found in literature (auto-biographic notes and contemporary descriptions) for a wide group of famous people. However, a definite diagnosis of epilepsy cannot be made in these cases. They include (Schneble 2003)

**Philosophers**  Socrates, Blaise Pascal and Friedrich Nietzsche

**Writers:**  Dante Alighieri, Francesco Petrarca, Molière, Guy de Maupassant, Lewis Carrol and Roald Dahl

**Musicians:**  Georg Friedrich Händel and Richard Wagner

**Scientists:**  Paracelsus, Isaac Newton and Hermann von Helmholtz

**Statesmen:**  Alexander the Great, Caligula and Lenin

**Church representatives:**  Hildegard von Bingen, Jeanne d'Arc and Martin Luther.

## 1.3  Outline

This document is divided into three parts: Introduction, Materials and Methods, Results and Discussion. This standard structure should guide the reader through the document in a clear way.

The first part is dedicated to background information on the topic. It consists of three chapters: The current chapter provides a short historical overview of epilepsy as well as a general introduction. As this study is highly interdisciplinary, Chapter 2 is devoted to the medical background of epilepsy and Chapter 3 to the technical background of the employed methodology.

The second part contains the methodological core of this study. Chapter 4 provides an overview of quantitative state-of-the-art approaches to epileptic seizure propagation analysis and outlines the novel framework proposed in this study. The chapter is rounded off by a presentation of the analyzed ECoG data. The following chapters detail the four technical methods which are combined in the seizure propagation analysis framework: Chapter 5 is concerned with the automatic detection of high frequency oscillations, Chapters 6 and 7 with a causal analysis of rhythmic epileptiform activity and Chapter 8 with a segmentation and clinically inspired classification of these rhythmic patterns.

The third part comprises two chapters. Chapter 9 compiles the findings of the framework for epileptic seizure propagation. Comparing the results of the individual methodological chapters with complementary medical findings allows to deduce the potential seizure onset zone as well as the direction of initial seizure spread. Chapter 10 discusses the particular situation of the analyzed patient as well as framework-related aspects (limitation to ECoG data, performance issues). It is concluded by a short outlook.

## 1.4 Notation

The technical notation employed throughout this document follows the one of the signal processing community (apart from the symbol of the complex unit, which is denoted by $i$ like in mathematical text books, not by $j$): Namely, $f$ symbolizes the non-normalized frequency in Hz and $H(f)$ a filter frequency response. Furthermore, we use square brackets $[\cdot]$ to indicate discrete arguments and round brackets $(\cdot)$ for continuous ones; $\cdot^T$ symbolizes transpose, $\cdot^*$ conjugate, $\cdot^H$ conjugate-transpose and $\hat{\cdot}$ estimation. As usually, bold capital letters denote matrices, bold lowercase ones vectors. For better distinction from variables, expectation is denoted by $\mathbb{E}\{\cdot\}$ and variance by $\mathbb{V}\{\cdot\}$. $\star$ denotes convolution, $\mathfrak{F}\{\cdot\}$ the discrete-time Fourier transform (DTFT) and $\mathfrak{F}^{-1}\{\cdot\}$ its inverse.

Throughout this study we will consider real-valued, stationary multivariate stochastic signals ($\mathbf{x}[n], n \in \mathbb{Z}$) with time index $n$, whose realizations represent ECoG recordings uniformly sampled at a sampling frequency $f_s$. With regard to the nature of such a signal, we will call each of its $K$ components $x_k[n]$, $k = 1, \ldots, K$, a *channel*. Consequently, we refer to $\mathbf{x}[n]$ as *multi-channel signal*. Its power spectral density (PSD) is denoted by $S_\mathbf{x}(f)$.

The epilepsy terminology follows the current guideline of the ILAE (International League Against Epilepsy) published by Berg et al. (2010). For electroencephalographic (EEG) expressions we stick to Noachtar et al. (1999)[4].

---

[4]German-speaking readers might be interested in the translation by Noachtar et al. (2004).

# 2 Medical Background

We assume that the reader is familiar with basic medical vocabulary, we refer to Stedman (2005)[1] for a medical dictionary.

## 2.1 Human brain

> Εἰδέναι δὲ χρὴ τοὺς ἀνθρώπους, ὅτι ἐξ οὐδενὸς ἡμῖν αἱ ἡδοναὶ γίνονται καὶ αἱ εὐφροσύναι καὶ γέλωτες καὶ παιδιαὶ ἢ ἐντεῦθεν, καὶ λῦπαι καὶ ἀνίαι καὶ υσφροσύναι καὶ κλαυθμοί.
>
> — Ἱπποκράτης: *Περὶ ἱερῆς νούσου*
>
> And men ought to know that from nothing else but (from the brain) come joys, delights, laughter and sports, and sorrows, griefs, despondency, and lamentations.
>
> — *On the Sacred Disease*[2] by Hippocrates, father of medicine

This section gives a brief overview of anatomy and physiology of the human brain. Further details can be found in e.g. Kandel et al. (2012).

Note that we will use the English expression rather than the Latin ones throughout this docuement; however the Latin or Greek terms (if different from the technical term used in English) are specified in italic between brackets in this section whenever a structure is mentioned for the first time.

### 2.1.1 Anatomy

We start with a top-level view on the human head. The skin *(cutis)* is the outermost coating of the human head *(caput)*. It covers the skullcap

---

[1] German-speaking readers might be interested in Grossmann (2013).
[2] Opening of Section 14; English translation by Francis Adams.
  Original text and English translation are accessible online via the PERSEUS project at www.perseus.tufts.edu (»de morbo sacro«).

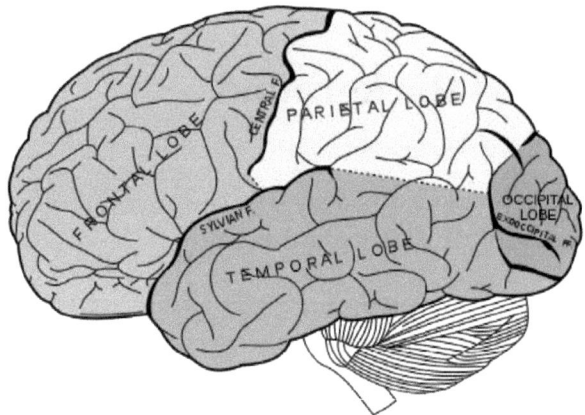

**Figure 2.1.1:** *Schematic representation of the cerebrum.* Principal fissures and lobes of the cerebrum are viewed laterally. Picture is in the public domain and taken from Wikimedia Commons (commons.wikimedia.org).

*(calvaria)* which is the upper part of the cranium, the latter surrounding the cranial cavity (which contains the brain).[3] Below the skullcap we find the meninges, a system of membranes consisting of three layers which protect the central nervous system: directly below the skull the dura mater *(dura mater cranialis)*, as middle layer the arachnoid mater *(arachnoidea mater cranialis)* and finally the pia mater *(pia mater cranialis)*. Between the two last we have the subarachnoidal space *(spatium subarachnoideum)* which is filled with cerebrospinal fluid, acting as a protective buffer.

The meninges cover the surface of the brain *(encephalon)*, i.e. the cortex, which consists of gyri and fissures *(sulci)*. The gray matter of the cortex is built up of neurons and covers the white matter situated below, which mostly contains glial cells and myelinated axons.

The cerebrum consists of two hemispheres (right and left), separated by the inter-hemispheric fissure (or longitudinal fissure, *fissura longitudinalis*). As Fig. 2.1.1 details, each of the two hemispheres is clustered into four lobes *(lobi)* which are separated by fissures:

---

[3] The skullcap is made up of the frontal, occipital, right and left parietal, right and left temporal, sphenoid, and ethmoid bones; therefrom the nomenclature of the lobes.

**Fontal lobe** *(lobus frontalis)*, bordered by the central fissure, the lateral fissure (also termed Sylvian fissure) and the cingulate fissure

**Temporal lobe** *(lobus temporalis)*, bordered by the lateral fissure and the preoccipital notch *(incisura preoccipitalis)*

**Parietal lobe** *(lobus parietalis)*, bordered by the central fissure, the lateral fissure, the cingulate fissure and the parieto-occipital fissure

**Occipital lobe** *(lobus occipitalis)*, bordered by the preoccipital notch and the parieto-occipital fissure

In addition, the limbic lobe *(lobus limbicus)* and the insular cortex (*lobus insularis*) are part of the cerebrum. The insular cortex is covered by the opercula (»little lids«) of the frontal, parietal and temporal lobes (thus not visible in Fig. 2.1.1).

The primary fissures of the cortex which separate the lobes are

**Central fissure** (*sulcus centralis*), located between frontal and parietal lobe; thus separates the (motoric) precentral gyrus and the (sensitive) postcentral gyrus.

**Lateral fissure** (*sulcus lateralis*), also called Sylvian fissure, separates frontal, parietal and temporal lobe; below this fissure, the lateral fossa (*fossa lateralis*) and the insular cortex are located.

**Parietooccipital fissure** (*sulcus parieto-occipitalis*), runs from the interhemispheric fissure along the medial hemispheric surface to the calcarine fissure; separates parietal and occipital lobe.

**Calcarine fissure** (*sulcus calcarinus*) located on the medial hemispheric surface like the parieto-occipital fissure.

**Cingulate fissure** (*sulcus cinguli*) separates the limbic lobe and the frontal and parietal lobes.

We refer the reader to an anatomical atlas, like e.g. the one by Paulsen and Waschke (2010), for detailed illustrations of the cerebral lobes and fissures.

## 2.1.2 Physiology

As already mentioned in the last subsection, the brain is primarily built up of two groups of cells: neurons and glial cells. The latter perform a number of critical functions, including structural and metabolic support as well as insulation. Although they were believed not to take part in active neural information transmission for a long time, this view has changed recently (Gourine et al. 2010). However, the exact mechanism remain unclear.

On the other hand, the role of neurons in information transmission is well understood. They belong to a group of specialized cells characterized by electrical excitability and the presence of synapses. As is well known, neural signal transmission happens electro-chemically, compare Fig. 2.1.2: Synaptic, i.e. chemical, signals from other neurons are received by the soma and dendrites. This excitation results, if above a threshold, in a raised action potential, compare the model of Hodgkin and Huxley (1952). Hereby, the excitability of the neuron is influenced by voltage-dependent ion channels (e.g. Na, K, Ca). The action potential then travels along the axon to its synaptic end where neurotransmitters are released.

These neurotransmitters open or close ionic channels of the subsequent neuron. Hereby two main types of transmitters act as antagonists: glutamate in an excitatory and GABA in an inhibitory way. The signal is thus transmitted chemically across the postsynaptic cleft and registered by receptors on the soma or dendrites of the next neuron. These receptors contain ionic channels themselves or act on other channels via »second messengers«. In either way, the chemical signal is re-converted into an electric stimulus which can provoke an action potential again.

The action potential itself is extremely short in duration (1-2 ms), while the postsynaptic potentials are considerably slower. Unlike action potentials, they do not follow the all-or-none law, but superpose each other. Thus, they

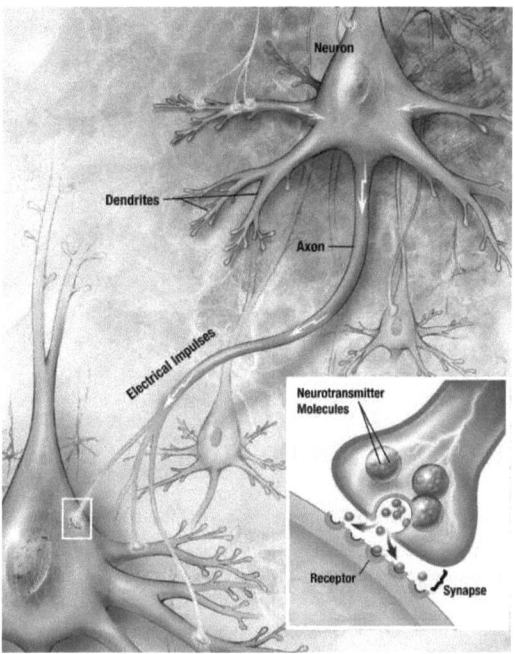

**Figure 2.1.2:** *Schematic representation of neural information transmission. Electrical transmission through action potentials along the axon and chemical transmission across the synaptic cleft by neurotransmitters. Picture is in the pubic domain and taken from Wikimedia Commons (commons.wikimedia.org).*

generate slowly changing sum potentials with wide range. This subcortically generated activity propagates to the surface, where it can be measured as rhythmic activity, e.g. by the EEG (compare Section 2.2).

The exact genesis of this rhythmic activity has not been clarified yet entirely. In principle there are two distinct mechanisms:

**Pulse generator** A strong pulse generator initially starts rhythmic activity and forces other neurons into its rhythm. Hereby one single neuron could already act as generator.

**Feedback processes** Several neurons participate in the genesis of rhythmic activity. Here all neurons of the network contribute equally, and the rhythmicity is a result of feedback processes.

The function of this rhythmic activity is widely unknown, except of some particular cases (e.g. gamma activity as support for neuronal coordination, »feature binding«).

While in physiological conditions the neural firing happens in a controlled way (leading to distinct synchronized areas), it is out of order in epilepsy patients during seizures (see Section 2.3). In the initial seconds of an epileptic seizures the pulse generator hypothesis serves as explanatory model, but in later stages the neuronal activity seems to be described better by feedback processes (Dudek et al. 1999).

## 2.2 Electroencephalography

> Wir sehen im Elektrenkephalogramm eine Begleiterscheinung der ständigen Nervenvorgänge, die im Gehirn stattfinden, genau wie das Elektrokardiogramm eine Begleiterscheinung der Kontraktionen der einzelnen Herzabschnitte darstellt.
>
> — Hans Berger: *Über das Elektroenkephalogramm des Menschen*
>
> In the electroencephalogram we observe an accompanying phenomenon of permanent nerve processes that take place in the brain, just like the electrocardiogram represents an accompanying phenomenon of contractions of individual heart segments.[4]
>
> — *On the Electroencephalogram of Man* by Berger (1929), pioneer in human electroencephalography

This section gives a short introduction to the field of electroencephalography. For further information we refer to two standard works, Niedermeyer and Lopes da Silva (1993) and Ebersole and Pedley (2003). Lüders and Noachtar (1994) is an EEG atlas with many recording examples.

---

[4]Own translation.

## 2.2.1 Definition

The *electroencephalography* (EEG, from the Greek terms εν = in, χεφαλή = head and γράφειν = to write) is a clinical diagnostic method for measuring the electrical activity of the cortex. This is done by recording of potential differences between electrodes placed on the scalp. Each of the measured potentials is the superposition of potentials caused by the electrical activity of the individual neurons located in the neighborhood of the respective electrode, compare Subsection 2.1.2. Recording of these potential differences over time yields the oscillating EEG signal. The *electroencephalogram* (abbreviated to EEG as well) is the graphical illustration of these oscillations. By convention, negative values of the potential difference are plotted on the positive ordinate, positive values on the negative ordinate (which is inverse to our everyday representation).

Nearly 100 years ago the German neurologist Berger (1929) successfully accomplished the task of recording and displaying human cerebral activity for the first time. Since then the EEG has been established as a standard non-invasive and cheap diagnostic tool in neurology, in particular in epilepsy diagnosis (compare Subsection 2.3.3).

Recorded cerebral activity shows oscillations with varying frequency, depending on anatomy and neurophysiology. The following frequency bands are typically distinguished:

**Delta ($\delta$):** < 4 Hz, e.g. in adults in slow-wave sleep (sleep stage N3).

**Theta ($\theta$):** 4-8 Hz, e.g. a specific form of epileptic discharges, see Subsection 2.3.6.

**Alpha ($\alpha$):** 8-13 Hz, e.g. posterior basic rhythm in adults.

**Beta ($\beta$):** 13-30 Hz, e.g. increased frontal beta activity due to benzodiazepine administration.

**Gamma ($\gamma$):** > 30 Hz, with 2 specific sub-bands: *ripple* band (80-250 Hz) and *fast ripple* band (250-500 Hz), see Subsection 2.3.6.

## 2.2.2 Recording conventions

In order to achieve comparability and reproducibility of electroencephalograms, Jasper (1958) proposed the *10-20-system* which still represents the recording standard today. The name refers to the fact that the distances between adjacent electrodes are either 10% or 20% of the distance from nasion to inion or between the left and right preaurical points, as shown in Fig. 2.2.1 (a).

Each electrode position has a letter to identify the lobe and a number to identify the hemisphere location. The letters F, T, P and O symbolize the respective lobes, the letter C is used for electrodes placed on the mid-line. Even numbers are assigned to electrode positions on the right hemisphere, odd ones to positions on the left hemisphere.

The 10-20-system can be extended to the so-called *10-10-system* (Chatrain et al. 1985), where all adjacent electrodes are separated by 10% of the reference distances on the skull, compare Fig. 2.2.1 (b). Note that four electrodes (marked in black color in the picture) have different labels in this system.

The advantage of this extended recording system lies in the increased spatial resolution. On the other hand, the increased effort for properly applying all electrodes is a drawback. Furthermore, in many applications the reduced number of EEG channels provided by the 10-20-system is sufficient. Depending on the context, a practical alternative is usage of the 10-20-system with some additional positions taken from the 10-10-system. For instance, in case of temporal lobe epilepsy additional usage of electrodes FT9/10 and TP9/10 might be of interest (as done in long-term video-EEG monitoring at Neurological Center Rosenhügel, see Subsection 2.3.6).

As the EEG represents potential differences rather than absolute values (compare Subsection 2.2.1), the electroencephalogram depends on the setup, i.e. how these differences are calculated. The most natural setup is the *reference setup*. Here one specific electrode is designated as common reference,

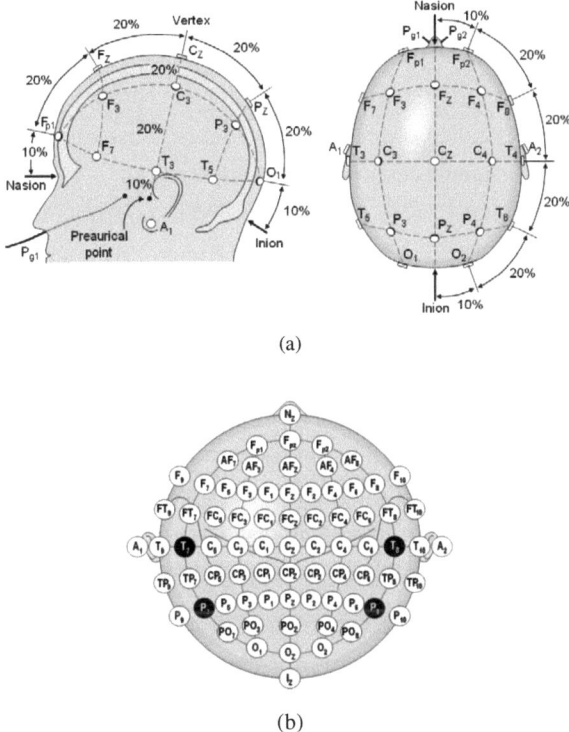

**Figure 2.2.1:** *EEG recording conventions. (a) standard 10-20-system used for routine EEG recording, (b) enhanced 10-10-system. Electrodes marked in black color have different labels in the 10-10 system. Pictures are taken from from the web edition (www.bem.fi/book) of Malmivuo and Plonsey (1995) whose material is free for publishing.*

and all other channels represent the respective potential differences to this electrode. Another possible way of recording is the *bipolar setup*. Here, differences are measured subsequently along defined paths between adjacent electrodes. Depending on the paths, one obtains various bipolar setups, including the prominent *longitudinal bipolar setup* (»double banana«: two paths from anterior to posterior on each hemisphere in banana-shape).

Note that nowadays, as EEG is registered by digital means, the setup plays a subordinate role in the recording process. This is due to the fact that, once the EEG has been recorded in reference setup, any other setup configuration can be calculated: The new channels are obtained by simply subtracting the desired reference from the actual channels. This allows the definition of the so-called *common average setup*, where the mean of all (or a selected set of) channels acts as virtual reference.

Therefore, digital EEG is simply recorded in reference setup, and the EEG program then displayed the traces in the appropriate setup on demand. Fig 2.2.2 shows exemplary EEG recordings in different setups: in bipolar setup in Fig. 2.2.2 (a) and in common average setup in 2.2.2 (b). Data are taken from a patient undergoing long-term video-EEG monitoring at Neurological center Rosenhügel in March 2011. The patient suffers from epilepsy with cavernoma in the right temporal lobe[5]. Note the prominent spike at electrode FT10 on the right hemisphere which is clearly visible in both setups. The method of localizing discharges in EEG (electrode FT10 in this example) will be discussed in the next subsection.

Digital EEG has usually been recorded at a sampling frequency of 128 Hz or 256 Hz. As the hardware (recording boxes, storage systems, processing power of computers) has become increasingly powerful in the last years, nowadays higher sampling frequencies such as 1024 Hz are also common. These allow, for instance, for the recording of oscillations in high-frequency bands (HFOs, compare Subsection 2.3.5).

### 2.2.3 Localization of discharges

As motivated in the last section, localization of discharges is needed for the determination of the position of EEG correlates (lateralization or even exact electrode position, if possible).

In principle, we can distinguish several levels of localization accuracy:

**focal or multi-focal:** only applies to invasive EEG (with increased spatial resolution), compare Subsection 2.3.6.

---

[5]See Subsection 2.3.2 for epilepsy classification.

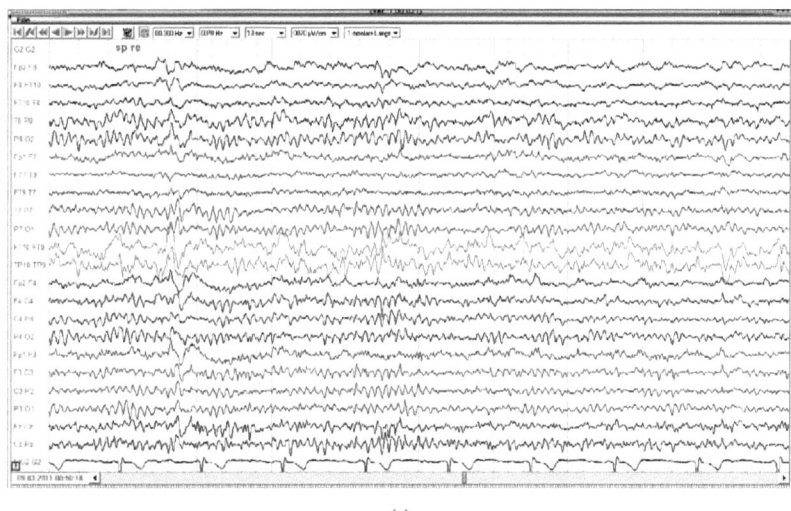

(a)

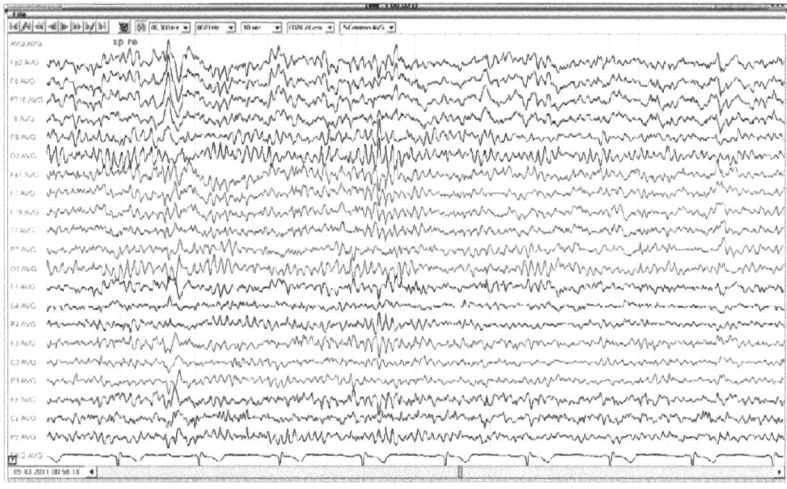

(b)

**Figure 2.2.2: *Exemplary EEG recordings.*** *(a) bipolar longitudinal vs (b) common average setup. In both cases a spike is clearly visible at electrode FT10 on the right hemisphere.*

**regional:** The discharges are limited to one lobe or to a part of one lobe.

**multi-regional:** The discharges affect at least 3 lobes on both hemispheres.

**lateralized:** The discharges are limited to one hemisphere, but a better spatial resolution is not possible.

**generalized:** Both hemispheres are involved.

In order to localize electroclinical correlates one has to consider the polarity convention and localization rules. In the following we give a very brief overview:

First, recall that the EEG is recorded by a differential amplifier with two input channels, say input 1 and input 2. It amplifies the potential difference input1 - input2. This leads to the following *polarity convention*:

**»up« rule:** We obtain a negative signal (displayed on the positive ordinate, thus »up«) if input 1 is negative or input 2 positive.

**»down« rule:** In analogy, we obtain a positive signal (displayed on the negative ordinate, thus »down«) if input 1 is positive or input 2 negative.

Second, the application of this polarity convention to the field potential (as distributed over the cortex) results in *localization rules*, which are summarized in Table 2.2.1. They are dependent on the setup, i.e. differ between bipolar and reference setup, where in the latter case the reference can be an electrode or a computed reference (e.g. common average, compare Subsection 2.2.2). The two important cases commonly encountered are set in bold-face type in Table 2.2.1: In case of bipolar setup a phase reversal indicates the maximum of the field potential, in case of reference setup and absence of phase reversal the maximal amplitude indicates the maximum of the field potential.

Consider the exemplary EEG recordings in Fig. 2.2.2 as an example: The electroclinical correlate of interest is the spike/sharp wave, an epilepsy-typical potential which is characterized by a negative field potential. In the

| Setup | Phase reversal | Localization |
|---|---|---|
| bipolar | no | min/max of field potential is located at the end of the electrode series |
| **bipolar** | **yes** | **min/max of field potential is located at the electrode of phase reversal** |
| **reference** | **no** | **reference electrode is min/max of field potential if min (i.e. reference properly chosen): channel with highest amplitude is max of field potential** |
| reference | yes | reference electrode is neither min nor max of potential field, thus located within field (i.e. reference not properly chosen) |

Table 2.2.1: *Localization rules in EEG for bipolar and reference setup. In either case, the description of the situation most commonly encountered is set in bold-face type.*

following we examine the (lateral) electrode series Fp2 - F8 - FT10 - T8 - P8 - 02 on the right hemisphere. As mentioned in Subsection 2.2.2, it combines electrodes from the 10-20 and the 10-10 system.

The bipolar setup in Fig. 2.2.2 (a) shows a negative phase reversal between F8-FT10 and FT10-T8, i.e. F8-FT10 and FT10-T8 point to (not from) each other. The phase reversal is therefore located at electrode FT10, which represents the maximal (negative) field potential. Thus, the spike/sharp wave is located at electrode FT10.

The common average setup in Fig. 2.2.2 (b) does not reveal any phase reversal, and the maximal amplitude is shown over electrode FT10. Therefore, FT10 represents the maximal field potential, the spike/sharp wave is thus located at FT10.

## 2.2.4 Artifacts

Artifacts are oscillations in the EEG which are not generated by cortical activity. They compromise the quality of the recordings, thus may influence the judgment of the clinician. In the worst case, this might even lead to misinterpretation of the EEG correlates.

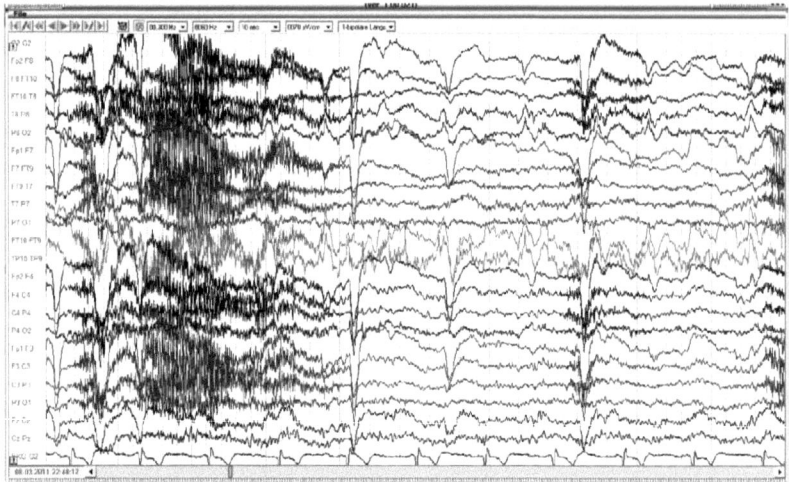

**Figure 2.2.3:** *Typical physiological artifacts.* Exemplary EEG recordings in bipolar longitudinal setup from a postictal phase revealing blink artifacts (frontal electrodes Fp1 and Fp2) and muscle artifacts (blockwise on all electrodes).

Artifacts are clustered into two groups, physiological artifacts and technical artifacts.

Physiological artifacts are caused by the patient. They are characterized by a typical field and a typical form of the EEG correlates and show spontaneous onset and offset. Typical physiological artifacts include

**blink artifacts** caused by the bulb of the eye which generates a dipole (positive at the cornea). When the lid is closed, the bulb quickly turns up, therefrom the artifact. Involved electrodes are (at least) Fp1/Fp2.

**eye movement artifacts** caused by (slower) lateral or vertical eye movements. Again, the frontal electrodes are involved.

**muscle artifacts** induced by muscle contractions (tensing of the temple muscle due to stress, motor activities during epileptic seizures, chewing). The EEG signal is blurred by oscillations of high frequency, typically in the $\beta$-band.

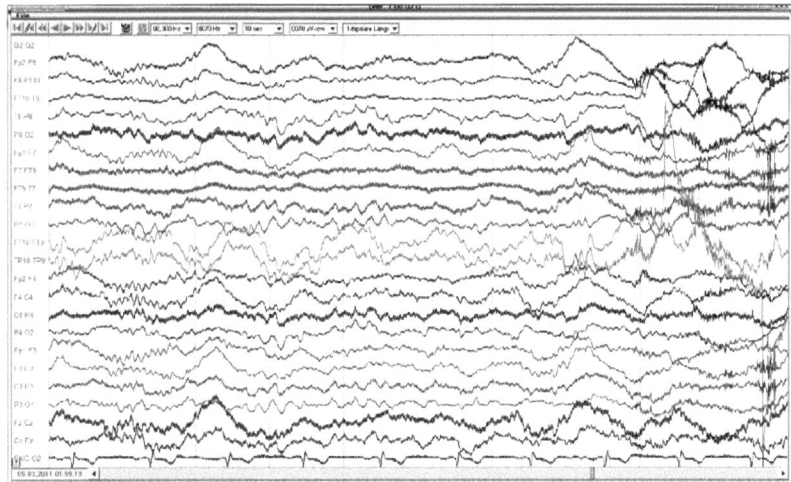

**Figure 2.2.4:** *Typical technical artifacts. Exemplary EEG recordings in bipolar longitudinal setup revealing line interference (all electrodes) and cable artifacts from loose electrodes (TP9-10).*

**ECG artifacts** caused by the dipole of the beating heart (top of the heart is positive). In order to facilitate the recognition of the sharp, low-amplitude artifacts, the ECG is recorded in parallel to the EEG.

**pulsation artifacts** caused by electrodes which are placed directly over an artery. Pulsation leads to a minimal electrode movement which generates these artifacts. They are again easy to recognize due to the parallel ECG trace, as they follow the heart beat by 200-300 ms.

**glossocinetic artifacts** caused by tongue movements as the tongue also constitutes a dipole (front end negative).

**galvanic skin artifacts** caused by sweating which alters the skin resistance. This results in a high-amplitude slow drift of the concerned channel.

Fig. 2.2.3 shows EEG recordings obtained during a post-ictal phase in the course of long-term video-EEG monitoring (see Subsection 2.3.6) at Neurological Center Rosenhügel. These recordings exhibit two prominent physiological artifacts, muscle artifacts (present block-wise on all channels) and blink artifacts (frontal electrodes Fp1 and Fp2 with a decreasing field in posterior direction).

On the other hand, technical artifacts are not directly caused by the patient, but by the instrumentation itself. Typical representatives of this group include

**line interference artifacts** caused by electromagnetic induction of the line frequency (50 Hz in Europe, 60 Hz in the US) which leads to characteristically blurred signals. Note that this artifact can be easily filtered out by setting a notch, i.e. a narrow band-stop, filter at the disturbing frequency.

**electrostatic induction artifacts** caused by electrostatic induction of instrumentation or the persons involved, e.g. by movement of the EEG assistant (amplified by inappropriate clothing). Electrostatic induction artifacts typically result in sharp (vertical) transients due to the locally induced potential with a slow decay back to the baseline level.

**electrode or cable artifacts** caused by the electrode itself (defect, dry or loose due to e.g. motor movement of the patient) or a broken cable. This type of artifact is easy to recognize as only the concerned channel alone shows irregularities (no field potential visible).

Fig. 2.2.4 shows EEG recordings from the same examination as before in Fig. 2.2.3, but during an interictal period. Here they exhibit two prominent technical artifacts, 50 Hz line interference (present on all channels) and electrode/cable artifacts (loose electrodes TP9/10).

## 2.2.5 Invasive EEG

EEG is a cheap and non-invasive diagnostic technique, but suffers from two main limitations: First, the spatial resolution of surface EEG is limited to a regional localization (compare Subsection 2.2.3). Second, EEG recordings are often impaired by artifacts (compare Subsection 2.2.4). In order to increase the spatial resolution and to reduce artifacts, in particular muscle artifacts, one can switch to invasive EEG.

In invasive or intracranial electroencephalography, the electrodes are implanted under the cranium, therefrom the name. There are basically two types of invasive EEG:

**Subdural EEG:** Subdural strip or grid electrodes are implanted under the dura mater, thus directly record cerebral activity from the cortex. This recording technique is therefor termed *electrocorticography (ECoG)*. Usage of strip electrodes only requires a minimally surgical intervention (insertion of the strip through a hole in the cranium and exact placement intracranially), but recording of activity is limited to the area along a line. On the other hand, grid electrodes (e.g. quadratic $8 \times 8$-grids) are able to record from a large area, but require a craniotomy for implantation. We refer to Quesney and Niedermeyer (1993) and Arroyo et al. (1993) for detailed information about ECoG.

**Intracerebral EEG:** Spenoidal or depth electrodes are pushed into the cerebrum, thus are able to directly record activity from deeper brain structures, e.g. the hippocampal region. The advantage of this technique is the increased spatial resolution of inner cerebral activity. However, the implantation of such electrodes requires increased preparatory effort in order to avoid damage of the neural vascular system.

A prominent application of invasive EEG is presurgical evaluation of epilepsy patients, compare Subsection 2.3.6. Localization is done in exactly the same way as in surface EEG (compare Subsection 2.2.3), as the recording technology itself, i.e. differential amplification, does not differ. Note, however, that there are no standard recording schemes in case of intracranial EEG, because each electrode implantation is tailored to the diagnostic needs of the individual patient.

## 2.3 Epilepsies

> Epilepsy is the name for occasional, sudden, excessive, rapid, and local discharge of grey matter.
>
> — *On the anatomical, physiological, and pathological investigation of epilepsies* by Hughlings-Jackson (1873), father of modern epileptology

This section gives an introduction to epilepsies, based on Baumgartner (2001) where we refer to for further details. Here we briefly discuss their definition, classification, diagnosis and therapy. The section is rounded off by an introduction to epilepsy surgery and a brief discussion of related social aspects.

### 2.3.1 Definitions

**Epileptic seizures** represent the clinical manifestation of excessive, hypersynchronous neuronal discharges of the cortex. They originate in a setting of both increased cell excitability and altered inter-neuronal synchronization (Tavee 2010). Clinical symptomatology hereby depends on the function of the affected neural network and can comprise disorders of advanced brain functions, limitation of consciousness, abnormal sensory or psychical perceptions, motor disorders or generalized spasms (see Subsection 2.3.2). An epileptic seizure is an unspecific reaction of the brain to various insults or lesions.

The time frame during an epileptic seizure is termed *ictal*, preceded by *preictal* and followed by *postictal* periods. The two latter are part of the *interictal* time frame, i.e. the one between two epileptic seizures.

**Epilepsies** form a heterogeneous group of neurological disorders which are characterized by unpredictable, unprovoked and recurrent epileptic seizures. The diagnosis of epilepsy depends on the persistence of the cause during interictal periods. Therefore, a single, unprovoked epileptic seizure does not lead to the diagnosis of epilepsy.

Note that acute symptomatic seizures (acute affection of the central nervous system, e.g. infections, traumata, cerebrovascular diseases, or acute systemic impairment, e.g. alcohol, sleep deprivation, medication, metabolic disorders) do not indicate epilepsy, as in this case an immediate activator can be identified. Neither do febrile seizures, which are, by definition, acute symptomatic seizures, but form their own subgroup due to their specific characteristics (age, genetic predisposition).

The ICD-10 classification of the WHO for epilepsies is G40.x (where x stands for the various syndromes, compare Subsection 2.3.2). According to the WHO, epilepsies are defined by two or more recurrent unprovoked epileptic seizures.

**Status epilepticus** (ICD-10 classification G41.x) is not in the scope of this study, but mentioned for the sake of completeness. According to the German Society for Neurology, the status epilepticus is

- either an epileptic seizure whose duration exceeds the duration of five minutes
- or a series of epileptic seizures which follow each other so rapidly that neither a clinical nor electroencephalographic restitution takes place interictally.

Epilepsies represent one of the most common neurological disorders, with a prevalence of 0.7% (Hirtz et al. 2007). Incidence shows two peaks, with one maximum in the first life months (with a strong decline after the first life year, a stable level in the first 10 years and a second decline in adolescence) and a second maximum in old age ($> 60$ years). Adulthood forms a minimum. Note that only 50% of epilepsies manifest before the age of 20, 25% after the age of 60 (Banerjee and Hauser 1997).

Although epilepsies comprise a wide group of various syndromes (compare Subsection 2.3.2), the course of disease follows of one of three paths in general:

**Complete remission** after an initial seizure, on a long-term perspective even without medication. This positive case is not in the focus of this study.

**Partial remission** is caused by remission and subsequent relapses, often in the course of discontinuation/reduction of medication. Compare Subsection 2.3.4 for references concerning medicinal therapy.

**No remission** is characterized by high seizure frequency with only short or even inexistent remissions in between. Compare Subsection 2.3.6 for an introduction to epilepsy surgery in case of therapy resistance.

Note that in about 30% of all cases a first seizure is followed by a second, i.e. develops into an epilepsy (Hauser et al. 1982).[6] Compare Schmidt and Sillanpää (2012) for patterns of remission in the history of treated epilepsy.

### 2.3.2 Classification

An important aspect in epileptologic classification is the difference between *syndromes* and *symptoms* (Noachtar et al. 1998). The (observed) epileptic seizure itself is the symptom of the disease. The description of the symptom is carried out by electroclinical analysis (visual inspection of the EEG correlates) and interpretation of the clinical symptomatology (semiology). The syndrome, on the other hand, depends on the etiology, i.e. the underlying type of cause. The diagnosis which syndrome the patient suffers from is a result of the synopsis of the findings (medical history, seizure semiology, EEG correlates, MRI findings, genetic data).

A particular symptom may be observed in different syndromes, and a particular syndrome may provoke different symptoms. Therefore, therapy recommendation is deduced from the syndrome, not from the symptom(s).

---

[6]Therefore, anti-epileptic medication is only advised after a first seizure if MRI or EEG is positive. Compare Subsection 2.3.4 for drug therapy.

| 1. Generalized seizures | 2. Focal seizures | 3. Unknown |
|---|---|---|
| 1.1 tonic-clonic<br><br>1.2 absence<br>— typical<br>— atypical<br>— absence with special features (myoclonic absence, eyelid myoclonia)<br><br>1.3 myoclonic<br>— myoclonic<br>— myoclonic atonic<br>— myoclonic tonic<br><br>1.4 clonic<br><br>1.5 tonic<br><br>1.6 atonic | 2.1 without impairment of consciousness or awareness<br>— with observable motor or autonomic components<br>i.e. »simple partial seizure« in old nomenclature<br>— involving subjective sensory or psychic phenomena only<br>i.e. »aura« in old nomenclature<br><br>2.2 with impairment of consciousness or awareness<br>i.e. »complex partial seizure« in old nomenclature<br><br>2.3 evolving to bilateral, convulsive seizure (involving tonic, clonic, or tonic and clonic components)<br>i.e. »secondarily generalized seizure« in old nomenclature | |

**Table 2.3.1:** *ILAE classification 2009 of seizures. The two groups of generalized and focal seizures are refined according to clinical symptomatology.*

In the following we summarize the ILAE classification 2009 (Berg et al. 2010) which defines the current scientific description of seizures, etiology and epilepsies.

First, seizures are clustered into three groups according to the electroclinical findings: generalized seizures, focal seizures and seizures of unknown type. The classification of generalized and focal seizures is refined based on the clinical symptomatology, as can be seen from Table 2.3.1.

Note that this classification is based on both the electroclinical findings and the clinical symptomatology. A classification purely with regard to semiology was proposed by Noachtar et al. (1998).

| |
|---|
| Electroclinical syndromes arranged by age at onset |
|    Neonatal period |
|    Infancy, e.g. West syndrome |
|    Childhood, e.g. Lennox-Gastaut syndrome |
|    Adolescence - Adult, e.g. juvenile myoclonic epilepsy |
|    Less specific age relationship |
| Distinctive constellations |
|    e.g. mesial temporal lobe epilepsy with hippocampal sclerosis |
| Epilepsies attributed to and organized by structural-metabolic causes |
|    Malformations of cortical development |
|    Neurocutaneous syndromes |
|    Tumor |
|    Infection |
|    Trauma |
| Angioma |
|    e.g. perinatal insults, stroke |
| Epilepsies of unknown cause |
| Conditions with epileptic seizures that are traditionally not diagnosed as a form of epilepsy per se |
|    Benign neonatal seizures |
|    Febrile seizures |

**Table 2.3.2:** *Summary of the ILAE classification 2009 of epilepsies. Table is shortened, only some prominent examples are shown.*

Second, etiology is grouped into three constellations:

**Genetic:** The epilepsy is the direct result of a known or presumed genetic defect. Verifiable by genetic tests (if existent) or by positive family history.

**Structural/metabolic:** A distinct structural or metabolic condition or disease causes the epilepsy. Structural lesions include acquired disorders such as stroke, trauma, infection. Verifiable by diagnostic imaging techniques (high-resolution MRI).

**Unknown cause:** The nature of the underlying cause is unknown (at the moment, but it may turn out to be e.g. a genetic defect)

Third, a summary of the syndrome classification is given in Table 2.3.2. Note that the classification point »Electroclinical syndromes arranged by age at onset« does not reflect etiology.

In this study we will focus on focal epilepsies, in particular on temporal lobe epilepsies (TLE) which are characterized by seizures originating from the temporal lobe. An excellent introduction to TLE can be found in Engel (1996a). Note, however, that terms like »temporal lobe epilepsy« or »frontal lobe epilepsy« are not supported by the current ILAE syndrome classification (compare Table 2.3.2), as these terms aim at a description of the symptoms. The syndrome »mesial temporal lobe epilepsy with hippocampal sclerosis« (»mTLE with HS«) represents an exception, as this is a distinctive constellation.

## 2.3.3 Diagnosis

The aim of diagnosis is to clarify whether the patient suffers from epilepsy, and if, to determine the syndrome and its etiology. This is a prerequisite for an appropriate therapeutic recommendation, compare Subsection 2.3.2.

This procedure involves the differential diagnosis, i.e. the active distinction of epilepsy from other neurological disorders with similar symptoms. In particular, epileptic seizures have to be delineated from psychogenic seizures (PNES, psychogenic non-epileptic seizures), convulsive syncopes and parasomnia.

In case of suspicion of epilepsy, the standard diagnosis involves

**Medical history** including family history, social anamnesis, seizure anamnesis (description of seizures by the patient and relatives, seizure frequency), general anamnesis of diseases, neurological disorders, epilepsy provoking factors.

**Neurological/psychiatric examination** in order to derive the neurological status and to assess potential cognitive or psychiatric symptoms.

**Electrophysiological examination** is a key step in the diagnosis, as the EEG is an important tool in epilepsy diagnosis (Smith 2005):[7] The aim of the 20-minute routine recording is to decide whether the patient's EEG shows interictal epilepsy-typical discharges (IEDs) or not. Such a prominent bio-marker (among many others) is the spike / sharp-wave which is a strong indication of epilepsy. Interictal spikes exhibit a rather good sensitivity (Walczak et al. 1992), but a very high specificity (Gregory et al. 1993). Therefore, occurrence of one single spike renders the EEG abnormal and confirms the diagnosis of epilepsy. In particular in temporal lobe epilepsies spikes are good lateralizing indicators (Janszky et al. 2001).

In order to increase the yield of epilepsy-typical potentials during these 20 minutes, so-called provocation methods are applied (photic stimulation with flashing light and hyperventilation). Thus, epilepsy can be diagnosed well by means of EEG even after the first seizure. However, it is important to consult the neurologist within 24 hours after the seizure (King et al. 1998). If the first routine EEG did not yield any results, a repetition of the examination is meaningful: The diagnostic value of repeated EEG recordings increases (Salinksy et al. 1987), but repetitions are only meaningful up to four times (Doppelbauer et al. 1993).

**Imaging techniques** like e.g. CT and mandatory high-resolution MRI complete the diagnosis, as positive findings allow to deduce a structural etiology (e.g. by high-resolution MRI). Compare Lai et al. (2010) for a recent review of neuro-imaging techniques in epilepsy. While high-resolution MRI is part of the standard diagnosis proce-

---

[7] A good overview of the use of EEG in epilepsy diagnosis is provided by Cascino (1996); Flink et al. (2002) give a general guideline for clinicians concerning the recording procedure.

dure, the use of PET, SPECT and fMRI is limited to presurgical evaluation (compare Subsection 2.3.6).

Advanced diagnosis is performed in the course of long-term video-EEG monitoring for presurgical evaluation (compare Subsection 2.3.6).

## 2.3.4 Therapy

This subsection gives a very brief overview of anti-epileptic therapy, we refer to Panayiotopoulos (2011) for a detailed introduction.

Drug therapy (anti-epileptic drugs, *AEDs*) is a cornerstone of modern therapy, with around 20 anti-epileptic agents currently in use. Compare the review articles by Das et al. (2012) and Porter et al. (2012) for their respective biochemical mechanisms and the one by Perucca and Tomson (2011) for indication, advantages, drawbacks, adverse events and dosage recommendation.

Many anti-epileptic agents tackle the two basic molecular mechanisms in epilepsy: the decrease of neuronal inhibition mediated by GABAergic channels (Macdonald 1997) and hyper-excitation due to excessive activation of the glutamatergic system (Dichter and Wilcox 1997). The mechanisms of many AEDs are well understood. For instance benzodiazepines like lorazepam, which is used as acute intravenous antiepileptic medication, causes increased inhibition via GABA-A receptors (Sieghart and Sperk 2002). On the other hand, the anti-epileptic agent lastly authorized in the EU, perampanel, acts as an antagonist of the glutamatergic AMPA receptor (Hanada et al. 2011). Most AEDs directly inhibit the sodium channels, e.g. carbamazepine. However, the exact biochemical mechanisms of the modern anti-epileptic agent levetiracetam have not been fully revealed yet.

Therapeutic approaches are derived from the diagnosis as the indication for different anti-epileptic agents depends on the respective syndrome. *Monotherapy*, i.e. administration of a single drug, stands at the beginning. In case of failure of the initial mono-therapy an alternative monotherapy is a valuable option: First, nearly 50% of newly diagnosed patients become seizure-free on the first anti-epileptic drug, with more than 90% of them at moderate dosing (Kwan and Brodie 2000, 2001); another 13% become seizure-free in alternative mono-therapy (Brodie et al. 2012). Second, administration of a single drug yields high efficacy and allows for minimization of adverse events and drug interactions, i.e. improved compliance.

In case of failure of monotherapies, *combination therapy* of two or more antiepileptic agents (*polytherapy*) is a subsequent step. The advantages and drawbacks of *rational polytherapy* are heavily discussed (French and Faught 2009, Kwan et al. 2011). On the one hand, one hopes on an additive effects of the individual agents. On the other hand, the efficacy of combinational therapy decreases exponentially with each additional anti-epileptic drug (Schiller and Najjar 2008). Moreover, the combination of several drugs increases the risk of adverse events.

Alternative options to medicinal therapy include epilepsy surgery (see Subsection 2.3.6), neuro stimulation (Vonck et al. 2012), in particular deep brain stimulation (Schulze-Bonhage 2009), and cetogenic diet (Caraballo and Vining 2012, Thammongkol et al. 2012).[8]

### 2.3.5 Localizing value of EEG

In the last subsections the goal was to confirm the diagnosis of epilepsy by means of EEG and to derive a proper medicinal therapy.

Here we are in a different setting: Surface EEG, and to a larger extent invasive EEG, provides information about origin and propagation of focal seizures. Now, the aim is not only to detect epilepsy-typical potentials, but to identify patterns with good localizing value. This information is needed for presurgical evaluation, compare Subsection 2.3.6.

---

[8]Potentially upcoming alternatives, which are still in an experimental stage in animals, include focal cooling (Fujii et al. 2012) or optical control of neural activity (Boyden et al. 2005).

### 2.3.5.1 Concept of the cortical zones

Lüders and Awad (1991) defined six cortical zones which play a role in the presurgical evaluation of candidates for epilepsy surgery, also compare the review by Rosenow and Lüders (2001).

**Symptomatogenic zone** is the region of the cortex that produces the ictal symptoms, if activated by an epileptiform discharge. It is, therefore, defined by analysis of the clinical signs.

**Irritative zone** is the region of the cortex that generates interictal epileptic discharges (IEDs), e.g. spikes, in the EEG. It is localized via EEG.

**Seizure onset zone (SOZ)** is the region where the clinical seizures originate from, i.e. where they are actually generated. It is usually the part of the irritative zone that generates spikes which are strong enough (e.g. repetitive spikes) for producing subsequent discharges. The SOZ is localized by surface or invasive EEG.

**Epileptogenic lesion** is a structural lesion that is causally related to the epilepsy (structural etiology, see Subsection 2.3.2). Localization is done by functional imaging, e.g. high-resolution MRI.

**Functional deficit zone** is the region of the cortex which is functionally abnormal during interictal periods. This dysfunction can either be a direct effect of the lesion or be functionally mediated (e.g. abnormal neuronal transmission). It is defined by neurological examinations, neuropsychological testing, EEG and radio-nuclear imaging.

**Epileptogenic zone** is the region of the cortex which is indispensable for generation of epileptic seizures. The epileptogenic zone is a theoretical construct in the context of epilepsy surgery: It may include an actual epileptogenic zone which generates the seizures before surgery (equivalent or smaller than the SOZ) and a potential epileptogenic zone which may generate seizures after surgery. Currently there is no diagnostic possibility for exact determination of the entire epileptogenic zone, because the existence of a potential epileptogenic zone only becomes apparent after surgery. The only possible conclusion is that, if the patient is seizure-free after surgery, the entire epileptogenic zone must have been situated within the resected area.

For our purposes we will only refer to the concept of the SOZ in this study.

### 2.3.5.2 Ictal rhythmic patterns

Ictal patterns provide a good localization of the SOZ of focal seizures, compare Foldvary et al. (2001): Rhythmic synchronous $\vartheta$-activity is the most frequent *pattern at onset* (in particular in case of TLE), followed by onset activity in the $\alpha$-range. Bare et al. (1994) also identified synchronous $\vartheta$-activity as most frequent ictal onset pattern in simple partial seizures. Moreover, Risinger et al. (1989) reported a good localizing value of rhythmic $\vartheta$-activity in seizures originating from the temporal lobe.

Kanemoto et al. (1997) found two different patterns at onset in TLE patients, both providing very good lateralization: an increase pattern with increasing amplitude in the $\vartheta$-range (recruiting $\vartheta$-rhythm again) and an decrease pattern with decreasing amplitude and frequency (irregular in the range of 2-5 Hz).

Note that the described rhythmic $\vartheta$-activity can be identified both in surface EEG and subdural recordings. However, the situation is slightly different in case of depth electrodes, where the morphology of the recorded activity heavily depends on the individual brain structure.

### 2.3.5.3 Interictal spikes

As the analysis/detection of spikes is not in the focus of this study, we only mention spikes for reasons of completeness.

Spikes define the irritative zone (see above), and their localizing value regarding the SOZ has been controversially discussed (Ebersole 1997):

While Blume et al. (1993, 2001) and Janszky et al. (2001) consider spikes to be useful in this type of analysis, Ray et al. (2007) recently pointed out that scalp spikes exhibit a poor localizing value. This is due to the propagation of the epileptic activity, compare e.g. the study by Baumgartner et al. (1995) which reveals propagation from the mesiobasal to the lateral temporal lobe.

### 2.3.5.4 High frequency oscillations

*High-frequency oscillations (HFOs)* represent an emerging research area with tremendously increasing interest in the last years. They occur in two frequency bands in the omega band above the standard EEG frequency range, as *ripples* (80-250 Hz) and *fast ripples* (250-500 Hz). Compare the reviews by Richardson (2011), Jiruska et al. (2010) and Bragin et al. (2010) as well as the one by Worrell et al. (2012) for recording techniques. HFOs can occur during interictal as well as ictal periods (Zijlmans et al. 2011). The huge interest in HFOs is due to their high clinical relevance, as they reveal an excellent localizing value for the SOZ in both cases.

On the one hand, the correlation of the interictal appearance of HFOs and the localization of the SOZ has been demonstrated by a number of studies (Jacobs et al. 2008, 2009, Brazdil et al. 2010) revealing a higher specificity and sensitivity as spikes. Moreover, HFOs could be bio-markers of the epileptogenic zone, compare the recent comprehensive review by Jacobs et al. (2012).

On the other hand, ictal HFOs are good markers of the SOZ, in particular in patients suffering from mTLE with HS (Usui et al. 2011)[9]. Ictal HFOs appear some time before conventional EEG onset, in average 8s prior (Khosravani et al. 2009) or 20s prior (Imamura et al. 2011).

HFOs are specific epilepsy-typical discharges due their high frequency and short duration. Initially they could only be detected by microelectrodes in intracranial EEG (Fisher et al. 1992, Alarcon et al. 1995). Then Jirsch

---

[9]HFOs were detected from unilateral medial temporal structures ipsilateral to HS. HFOs hardly ever showed contralateral propagation, and in case of bitemporal conventional EEG onset HFOs were detected on the side of HS.

et al. (2006) showed that HFOs can be recorded intracranially with commercially available macroelectrodes. Moreover, their detection has not been possible in surface EEG for a long time, as the skull acts as a low-pass filter. Surface EEG of children was an exception due to the thin infantile skull, compare e.g. the studies by Wu et al. (2008), Inoue et al. (2008) or Kobayashi et al. (2009). Only recently Andrade-Valenca et al. (2011) were able to detect HFOs in surface EEG of adults.

Recently, the appearance of ultra-high frequency oscillations above 500 Hz (Kobayashi et al. 2010) or even 1000 Hz (Usui et al. 2010) was reported in intracranial EEG.

HFOs often occur together with spikes, but their occurrence is independent of the one of the spikes. Urrestarazu et al. (2007) distinguish three different settings: occurrence of HFOs alone, HFOs on top of a spike (the most frequent case), HFOs being invisibly embedded into a spike (unless filtered). Spikes and HFOs have different patho-physiological mechanisms, the latter being closer to seizures. Unlike spikes (Gotman and Koffler 1989), HFOs do not increase after seizures, but decrease with medication (like seizures), see Zijlmans et al. (2009b).

We refer to Ozaki and Hashimoto (2011) for the physiology of HFOs, in particular to Jefferys et al. (2012) for the different physiological mechanisms of physiological and pathological HFOs. For instance, Wendling et al. (2002) provided a model explaining the transition from interictal to ictal fast activity by the impairment of dentritic inhibition.[10]

### 2.3.5.5 Ictal slow shifts

Moreover, a new class of biomarkers has become popular recently: *ictal slow shifts*. They are characterized by a slow negative baseline shift and correlate well with the SOZ, compare recent studies by e.g. Rodin and

---

[10]In general many psychiatric disorders lead to an increase of the $\gamma$-activity, in particular epilepsy (Herrmann and Demiralp 2005). Not only focal seizures reveal high-frequency activity, but also primarily generalized seizures lead to increased $\gamma$-activity (Willoughby et al. 2003).

Modur (2008) and Ren et al. (2011). In order to capture this infraslow activity, appropriate recording hardware is needed (Vanhatalo et al. 2005).

### 2.3.6 Epilepsy surgery

About 30% of patients suffer from therapy-resistant epilepsy (Schuele and Lüders 2008), i.e. persistent seizure freedom cannot be achieved by »adequate trials of two tolerated and appropriately chosen and used AED schedules (whether as monotherapies or in combination)« (Kwan et al. 2010). In these cases a resection of the epileptic tissue (i.e of the epileptogenic zone, compare Subsection 2.3.5) might render the patient seizure-free.

In this subsection we will briefly discuss the necessary presurgical evaluation, the surgical intervention and the post-surgical outcome, i.e. the quantified success of the surgery. We refer to the two standard works Engel (1996b) and Lüders (1991) for further information on epilepsy surgery.

#### 2.3.6.1 Presurgical evaluation

Therapy-resistant patients are admitted to the *Epilepsy Monitoring Unit (EMU)* of specialized neurological departments for prolonged presurgical evaluation. At such a center, the examination comprises long-term video-EEG monitoring, functional imaging and neuropsychological testing.

The goal of long-term video-EEG monitoring is to determine where seizures start from (detection of the SOZ) and where and how fast they propagate (determination of the seizure spread). Both dimensions are of clinical relevance: A clearly prescribed SOZ (located outside functionally indispensable regions) is a prerequisite for surgical intervention. Seizure spread has an impact on the post-surgical outcome, see below.

This analysis happens in two phases: First, trained specialists analyze the surface EEG recordings by visual inspection (*non-invasive phase*, Baumgartner and Pirker (2012)). In case of focal seizures, they mainly consider the ictal rhythmic synchronous activity, but also look at interictal spikes.

Compare Subsection 2.3.5 for the localizing value of EEG correlates regarding the SOZ. The video recorded in parallel allows for an analysis of the symptomatology, e.g. of lateralizing signs like version of the head or the »figure of four« (Rosenow and Lüders 2001, Rosenow et al. 2001). The semiology allows to draw conclusions on the symptomatogenic zone. Finally, high-resolution MRI can indicate a lesion, if existing.

However, surface EEG is not always sufficient for exact localization of the SOZ, e.g. lateralization can be derived, but no exact localization is possible (see Subsection 10.2.1 for an example). Moreover, if activity originates from deep structures like the hippocampus or the inter-hemispheric surface of the frontal lobe, surface EEG might not reveal any ictal potentials (Hashiguchi et al. 2007). In particular in case of mTLE, activity might propagate too quickly to both hemispheres in surface EEG (Lieb et al. 1976).

Thus, in order to increase spatial resolution (Behrens et al. 1994, Zumsteg and Wieser 2000) invasive EEG can be used in a second step (*invasive phase*, Kahane and Spencer (2012)). Subdural stripe/grid electrodes or depth electrodes are implanted according to the preliminary information obtained from surface EEG, compare Subsection 2.2.5.[11] For instance, Pacia and Ebersole (1997) exposed which patterns in deep brain structures correlate with surface potentials. However, even with these invasive techniques the SOZ cannot be localized adequately in about 20% of patients (Pondal-Sordo et al. 2007). Thus, these patients cannot be offered a surgical therapy and the electrodes have to be removed without resective surgery.

Despite these cases, several studies have proved that the increased effort of invasive EEG is justified, as it provides added value for the localization of the SOZ (see e.g. Guangming et al. (2009) for a recent study in case of TLE). Intracranial EEG is particularly useful for this purpose if EEG and MRI are not congruent, compare Pondal-Sordo et al. (2007), Henry et al. (1999) and Engel et al. (1981).

---

[11]Compare Lesser et al. (2010) for a review on subdural electrodes and Pacia and Ebersole (1999) for indications for the respective type of electrode.

Finally, intracranial EEG has another advantage: Highly-sampled invasive EEG gives access to the emerging field of high frequency oscillations, compare Subsection 2.3.5.

In an ideal case, the synopsis of all findings derived during the presurgical evaluation yields a conclusive picture of the potential epileptogenic zone (compare Subsection 2.3.5): long-term video-EEG monitoring (SOZ and symptomatogenic zone, irritative zone defined by interictal spiking), functional imaging in case of structural or metabolic deficits (epileptogenic lesion) and neuropsychological testing (functional deficit zone).

#### 2.3.6.2 Surgical intervention

The goal of the surgical intervention is removal of the seizure generating tissue, i.e. of the epileptogenic zone.

The precision of the surgical intervention has increased enormously since the first resective intervention by Penfield (1934) due to the technical possibilities both in presurgical diagnosis as well as in surgery itself. Bailey and Gibbs (1951) carried out the first *lobectomy* (resection of the temporal lobe in the reported case), and Wieser and Yasargil (1982) reported the first successful selective *amygdala-hippocampectomy*, the modern cornerstone of epilepsy surgery in mTLE with hippocampal sclerosis.

The use of intraoperative ECoG leads to even more target-oriented surgery. The recorded discharges provide information for the extent of the resection and post-surgical seizure control (Stefan et al. 2008).

Finally note that a study by Alarcon et al. (1997) underlines the usefulness of the concept of the epileptogenic zone in epilepsy surgery. It showed that removal of all »leading regions« (i.e. regions with the first peak in discharge, most commonly the hippocampus, the parahippocampal gyrus and the superior temporal gyrus) is important for a positive postsurgical outcome (see below), but resection of all discharging areas is not required.

| Class | Characterization |
|---|---|
| I | Free of disabling seizures: Seizure free or no more than a few early, nondisabling seizures; or seizures upon drug withdrawal only |
| II | Rare disabling seizures (»almost seizure-free«): Disabling seizures occur rarely during a period of at least 2 years; disabling seizures may have been more frequent soon after surgery; nocturnal seizures |
| III | Worthwhile improvement: Seizure reduction for prolonged periods but less than 2 years |
| IV | No worthwhile improvement: Some reduction, no reduction, or worsening are possible |

**Table 2.3.3:** *Classification of post-surgical outcome according to Engel (1996b). Surgical outcomes are rated from I (best) to IV (worst).*

### 2.3.6.3 Postsurgical outcome

The *post-surgical outcome*, i.e. the success of the surgical intervention regarding seizure frequency, is classified according to Engel (1996b), compare Table 2.3.3. The ultimate goal of any resection is to reach freedom from seizures at least under medication (class I), seizure freedom without medication is a plus. An alternative classification was proposed by the ILAE (Wiesner et al. 2001), which groups the post-surgical outcome into six classes.

The long-term seizure outcome varies significantly among the different syndromes, compare a recent meta-analysis of individual studies (Téllez-Zenteno et al. 2005): While temporal lobe resection promises a class I surgical outcome in 66% of all cases (other individual studies report similar results: Wiebe et al. (2001) 58%, Clusmann et al. (2002) 71%), occipital and parietal resection only lead to seizure freedom in 46% of patients. In case of frontal lobe resection the outcome decreases to 27% success rate.

An important aspect to consider is the potential influence factors the post-surgical outcome depends on. In case of TLE the outcome is better, if (Schulz et al. 2000) the patient has focal seizures without contralateral propagation, 100% unilateral IEDs and no asynchronicity between the hemispheres. Other studies confirm these findings: According to Lee et al. (2006), the postsurgical outcome of anterior temporal lobectomy in mTLE (mesial temporal sclerosis) is poor in case of contralateral propagation or bitemporal asynchronicity. Aull-Watschinger et al. (2008) report unilateral IEDs as highly significant for class I, and Clusmann et al. (2002) underline the importance of a single and lateralized focus. Furthermore the outcome is better if a structural lesion is revealed by MRI (Guldvogl et al. 1994), with best chances in case of tumors (Spencer 1996).

Besides conventional electroclinical correlates HFOs impact the post-surgical outcome, compare e.g. Jacobs et al. (2010), Wu et al. (2010), Akiyama et al. (2011): Removal of the HFO-generating tissue leads to good surgical outcome. Note, however, that the hypothesis does not hold that the more HFOs a tissue generates, the higher the resulting seizure frequency (Zijlmans et al. 2009a). Thus, seizure outcome does not depend on which HFO generating area has been removed, but only on the fact that the entire HFO-generating area has been resected.[12]

Finally, the speed of seizure propagation influences the post-surgical outcome. Lieb et al. (1986) showed, in a study based on depth electrodes, that the outcome is positively influenced, if the intracranial propagation from the epileptogenic hippocampus to the contralateral hippocampus lasted for more than 50 seconds; short propagation times smaller than 5 seconds contributed significantly to a poor outcome. Another ECoG-based study (Weinand et al. 1992) showed that an interhemispheric propagation time of at least 8 seconds was significant for good post-surgical outcome. These findings underline the clinical relevance of seizure propagation analysis.

---

[12]Therefore, Jacobs et al. (2010) conclude that HFOs might be potential bio-markers of the epileptogenic zone.

## 2.3.7 Social aspects

As is well known, the notion of health is not limited to the physical dimension:

> Health is a state of complete physical, mental and social well-being and not merely the absence of disease or infirmity.[13]

In case of epilepsy the mental and social dimensions are particularly important (Thorbecke and Pfäfflin 2012). Epilepsy patients have been stigmatized throughout centuries (compare Section 1.2), and even nowadays they are exposed to social pressure in developing countries like the ones of sub-Saharan Africa (Baskind and Birbeck 2005). Moreover, depression represents a frequent comorbidity in chronic epilepsy with approximately one third of patients suffering from major depression (Hermann et al. 2000). Many patients have to cancel their education or suffer from unemployment due to their restriction of activities in daily life.

Therefore, increased interest in a quantitative analysis of social aspects of anti-epileptic therapy has come up in the last years. In order to measure the patient's quality of life, Cramer et al. (1996) developed the QOLIE-10 questionnaire which has been extended, e.g. to the NEWQOL by Abetz et al. (2000). A more comprehensive set of questionnaires for quality of life measurement is the PESOS (May and Pfäfflin 2001). Part of patient's satisfaction is the absence or at least rare incidence of therapy-induced adverse events. The Liverpool Adverse Event Profile questionnaire proposed by Baker et al. (1994, 1997) measures the severity of common adverse events of anti-epileptic medication. Furthermore, Gilliam et al. (2006) worked out the NDDI-E questionnaire which aims at a rapid detection of depression in epilepsy patients.

---

[13] Preamble to the Constitution of the World Health Organization as adopted by the International Health Conference, New York, 19-22 June, 1946; signed on 22 July 1946 by the representatives of 61 States (Official Records of the World Health Organization, no. 2, p. 100) and entered into force on 7 April 1948.

# 3 Statistical Background

We assume that the reader is familiar with the basic notions of signal processing – we refer to Oppenheim and Schafer (1989) and Mitra (2002) for theory and applications of signal processing. In case the reader is interested in a more mathematical introduction to the topic of time series analysis, we recommend Brockwell and Davis (1991) and Lütkepohl (2007) or Hannan and Deistler (2012) for an in-depth theoretical view.

For a general background on EEG signal processing we refer to Tong and Thakor (2009) and Sanei and Chambers (2007). Varsavsky et al. (2011) provide additional background information on EEG signal processing in the field of epileptology, including physical models of EEG generation in the cortex.

## 3.1 Non-parametric spectral estimation

> It is convenient to have a word for some representation of a variable quantity which shall correspond to the 'spectrum' of a luminous radiation. I propose the word periodogram [...].
>
> — Schuster (1898): *On the Investigation of Hidden Periodicities with Application to a Supposed 26 Day Period of Meteorological Phenomena*, foundations of non-parametric spectral estimation

Spectral estimation deals with determining the characteristic frequency content of a signal, i.e. finding predominant frequencies or calculating the signal power in specific frequency bands. This is an important task in signal processing and has gained increasing importance in our digitized world. Nowadays its applications include such different areas as mobile communication, biomedical engineering or financial analysis.

The aim of this section is to quickly review the concept of non-parametric spectral estimation and to present one famous method in detail, the Welch power spectral estimation method. Parametric estimation is discussed in the context of autoregressive modeling, see Section 3.2.

This section is based on two classical text books of spectral estimation theory: Marple (1987) and Jenkins and Watts (1968). Hereby, the first provides an applied introduction from an engineering perspective including the presentation of all formulas in physical units (e.g. frequencies in Hz rather than in normalized rad), the second presents a mathematical formulation of the topic with technically rigorous proofs. Furthermore, the reader will find a good summary (though in German) of the main ideas and methods of non-parametric spectral estimation in the diploma thesis of Kilga (1993).

This section is limited to a univariate formulation for simplicity reasons. We refer to Chapters 15 and 16 of Marple (1987) for a multivariate extension of the spectral estimation methods as well as to The MathWorks, Inc. (2010) for details on their implementation in Matlab.

Finally note that we use physical frequency units ($f$ in Hz) instead of normalized ones ($\omega$ in rad). As is well known, they are related by

$$\omega = 2\pi f T,$$

where $T = f_s^{-1}$ denotes the inverse of the signal sampling frequency $f_s$. For a sequence $y[n]$ of finite energy, the pair of discrete-time Fourier transforms (DTFT) becomes[1]

$$Y(f) = \mathfrak{F}\{y[n]\} = T \sum_{n=-\infty}^{\infty} y[n] e^{-2i\pi f T n}$$

$$y[n] = \mathfrak{F}^{-1}\{Y(f)\} = \int_{-\frac{1}{2T}}^{\frac{1}{2T}} Y(f) e^{+2i\pi f T n}.$$

Note that in this section we will refer to Fourier-transformed signals by their respective capital letters, e.g. $Y(f) = \mathfrak{F}\{y[n]\}$ as above.

---

[1]The pre-multiplication with the inverse sampling frequency $T$ ensures a correct scaling, compare Section 2.8 of Marple (1987), »The issue of scaling for power determination«.

### 3.1.1 Typical issues

Often the question arises whether non-parametric or parametric estimation is more appropriate. Naturally occurring signals can be classified into one of the following two types (Mitra 2002):

**Noise-like random signals,** e.g. unvoiced speech signals like »f« or »s« are suitable for non-parametric spectral estimation. As an advantage, no information about an underlying model is needed.

**Signal-plus-noise random signals** e.g. seismic signals, EEG signals or similar ones can well be analyzed by employing parametric spectral estimation methods. However, this approach exceeds a »pure« estimation and involves a modeling step of the data, thus requires a-prior-knowledge about the signal-generating process.

Typical issues in spectral estimation involve (compare Chapter 1 of Marple (1987) for more details):

**Resolution** is an important property of a spectral estimator denoting its ability to distinguish distinct frequencies by separate peaks. Although this is an intuitive definition[2] and it somehow depends on a visual analysis of the output, we consider it to be sufficient for our purposes. Thus, for the sake of simplicity, we stick to this point of view and give an example as motivation. When applying an estimator with a high resolution to a superposition of two sinusoid with the respective frequencies $f_1$ and $f_2$, we would expect the following behavior: to show a flat line all over the spectrum except of two distinct peaks at frequencies $f_1$ and $f_2$. Each of them should be as »sharp«, i.e. as thin and high, as possible so that the two peaks are clearly circumscribed and do not overlap each other.
Spectral resolution will be of importance throughout this study.

---

[2] We refer to Chapters 2 and 5 of Marple (1987) for an exact discussion of resolution.

**Signal detectability** comes into play when spectral estimation is used in order to detect the presence of a signal (compare Section 3.5 on signal detection). In this case it is not of primary interest to distinguish two frequency peaks, but rather to assure that the estimation method does reliably indicate a certain frequency. Again, the evaluation of the result is subject to a visual analysis and subsequent interpretation of the output graph.

**Bias and variance** of the spectral estimator are the key statistical properties. The higher the variance, the »noisier« the graph of the estimated spectrum appears – the lower the variance, the »smoother« the graph looks. We always have to choose a trade-off between frequency resolution and variance for non-parametric estimation methods.

In this section we will be interested in the bias and the variance of the estimation method. This allows to determine whether the estimator is

- unbiased, i.e. the expectation of the estimator equals the parameter;
- consistent, i.e. the variance of the estimator converges to zero with increasing sample size and is asymptotically unbiased. This concept assures that an increasing sample size leads to a preciser estimation.

**Stationarity** of the first and second moments is a prerequisite for standard approaches in signal processing. Thus, the methods presented in the following are suitable for stationary signals. However, as many signals (such as speech or bio-signals) are highly non-stationary, one of the two following approaches is often chosen in practical applications.

- The first is to model the signal as *short-term stationary*, i.e. as a sequence of stationary regimes separated by abrupt changes (structural breaks). Here we assumes that sufficiently small segments of the signal show stationary properties. The idea is to cut the signal into a set of segments, estimate the spectrum in each of them and then calculate an »average spectrum«. A prominent application of this idea is the Welch spectral estimation method, see Subsection 3.1.3.

– A complementary approach is to assume that the statistical properties of the signal change in a continuous, but slow way. The appropriate methodology to cope with this behavior is *adaptive estimation*. By sliding along the time axis, the method constantly adapts its estimation to the new statistical regime. We refer to Priestley (1965) for the historical foundation of local stationarity as well as to Dahlhaus (1997) for a discussion of the statistical properties of the estimators.

## 3.1.2 Preliminary definitions

In the following we consider a univariate stationary signal $x[n]$ of mean zero.

Let $\hat{r}_x[m]$ denote the unbiased correlation estimator and $\check{r}_x[m]$ the biased correlation estimator. As is well known, they are closely related by

$$\check{r}_x[m] = \frac{N - |m|}{N} \hat{r}_x[m], \qquad (3.1.1)$$

with $N$ the sample size used for estimation.

In the following we assume that the correlation sequence is absolutely summable, i.e. decreases to zero with increasing lag,

$$\sum_{m=0}^{\infty} |r_x[m]| < \infty. \qquad (3.1.2)$$

Furthermore let $w_N^R[n]$ denote the rectangular window (of length N) and $w_N^{cR}[n]$ the centered rectangular window. The latter is shifted by $\frac{N}{2}$ on the time axis, i.e. we have

$$w_N^R[n] = w_N^{cR}\left[n - \frac{N}{2}\right]. \qquad (3.1.3)$$

We will often need a specific function:

**Definition 1** (Dirichlet kernel). The *Dirichlet kernel* $D_N(f)$, also denoted by *digital sinc*$^3$, is given by

$$D_N(f) = T \frac{\sin(N\pi fT)}{\sin(\pi fT)}. \qquad (3.1.4)$$

**Lemma 1.** *The DTFT of the centralized rectangular window is the Dirichlet kernel, i.e.*

$$W_N^{cR}(f) = D_N(f).$$

*Proof.* Appendix. □

**Corollary 1.** *The DTFT of the rectangular window is given by a phase-shifted Dirichlet kernel, i.e.*

$$W_N^R(f) = e^{-i\pi fTN} D_N(f).$$

*Proof.* Appendix. □

As it is well known, multiplication in the time domain and convolution in the frequency domain correspond to each other,

$$\mathfrak{F}\{w[m]\, r_x[m]\} = W(f) \star S_x(f), \qquad (3.1.5)$$

where $w[m]$ denotes an arbitrary window of finite energy in the time domain, $W(f)$ the DTFT of the window in the frequency domain and $S_x(f)$ the power spectral density (PSD) of the signal (see (3.1.6) in Subsection 3.1.3).

As the Dirichlet kernel shows considerable side lobes, the PSD of the windowed data will be »smeared« due to the convolution of the PSD of the raw data with the Dirichlet kernel $D_N(f)$. This behavior is termed *leakage effect*. Note that the length of the window controls the form of the frequency response. With increasing size of the window length the main lobe width decreases which leads to a higher frequency resolution of the spectral estimation. However, the computational complexity increases at the same time.

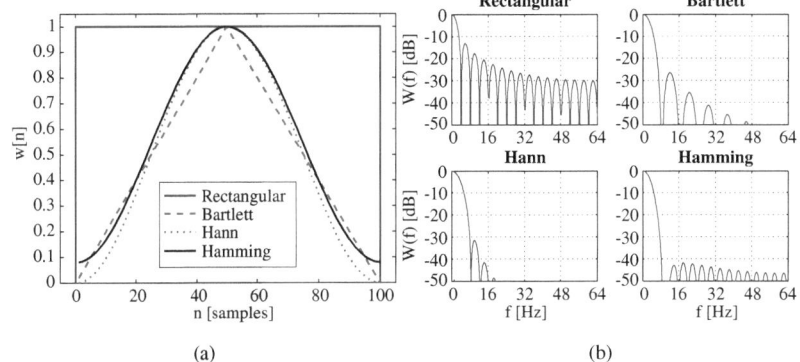

**Figure 3.1.1:** *Commonly used windows.* (a) Appearance of the rectangular, Bartlett, Hann and Hamming window in the time domain. (b) Respective gains in dB in the frequency domain.

| Window type | Peak side-lobe amplitude (relative) | Approximate width of main lobe |
|---|---|---|
| Rectangular | $-13\,\text{dB}$ | $\frac{4\pi}{N+1} \triangleq \frac{2}{T\cdot(N+1)} = 8\,\text{Hz}$ |
| Bartlett | $-25\,\text{dB}$ | $\frac{8\pi}{N} \triangleq \frac{4}{T\cdot N} = 16\,\text{Hz}$ |
| Hann | $-31\,\text{dB}$ | $\frac{8\pi}{N} \triangleq \frac{4}{T\cdot N} = 16\,\text{Hz}$ |
| Hamming | $-41\,\text{dB}$ | $\frac{8\pi}{N} \triangleq \frac{4}{T\cdot N} = 16\,\text{Hz}$ |

**Table 3.1.1:** *Characteristics of commonly used windows.* Main characteristics of the rectangular, Bartlett, Hann and Hamming window. Width of main lobe is given for a sampling frequency of 128 Hz.

In order reduce the leakage effect, one tends to find a trade-off between the magnitude of the side lobes and the width of the main lobe. For this purpose various windows have been suggested in literature which taper the data as smoothly as possible to zero at both ends of the window.[4] The variety reaches from simple triangular windows (the *Bartlett* window) to more complicated shapes based on functions of cosines (e.g. *Hann* window or *Hamming* window), compare Fig. 3.1.1 and Table 3.1.1 for their characteristics.

---

[3] Compare: The function *sinc* is defined by $\text{sinc}(f) \triangleq \frac{\sin(\pi x)}{\pi x}$.
[4] Therefore, windows are sometimes called *tapering functions*.

### 3.1.3 Indirect estimation methods

The power spectral density is commonly expressed as the discrete-time Fourier transform of the autocorrelation sequence[5]

$$S_x(f) = \mathfrak{F}\{r_x[m]\} = T \sum_{m=-\infty}^{\infty} r_x[m]\, e^{-2i\pi f T m}. \qquad (3.1.6)$$

One possible PSD estimator is therefore

**Definition 2.** The *correlogram with biased correlation* is given by

$$\check{S}_x^C(f) = T \sum_{m=-L}^{L} \check{r}_x[m]\, e^{-2i\pi f m T} \qquad (3.1.7)$$

with $\check{r}_{xx}[m]$ the biased correlation estimator, i.e.

$$\check{r}_x[m] = \begin{cases} \frac{1}{N} \sum_{n=0}^{N-m-1} x[n+m]\, x[n]^* & 0 \leq m \leq N-1 \\ \frac{1}{N} \sum_{n=0}^{N-|m|-1} x[n+|m|]^*\, x[n] & (N-1) \leq m \leq 0 \end{cases}$$

As the estimation of the autocorrelation has to be performed as an intermediate step, this estimation approach is termed *indirect*.

We will need the correlogram with biased correlation in the proofs of lemmata 3 and 4 in the next subsection. Some important properties are summarized in

**Lemma 2.** *Properties of the correlogram with biased correlation.*

1. The correlogram with biased correlation is a biased estimator of the PSD.

2. The correlogram with biased correlation is an asymptotically unbiased estimator of the PSD.

*Proof.* Appendix. □

---

[5]Its existence is assured by assumption (3.1.2).

Note that the PSD estimate (3.1.7) is the convolution of the true spectrum with a Dirichlet kernel, compare the proof of lemma 2. Therefore it is »smeared« due to the side lobes. In order to reduce the leakage effect, Blackmann and Tukey (1958) introduced a weighted version of the correlogram estimate, the Blackman-Tukey correlogram.

### 3.1.4 Direct estimation methods

#### 3.1.4.1 Periodogram

A well known PSD-estimator dates back to the end of the 19th century, when Schuster (1898) analyzed cyclic phenomena in nature.

**Definition 3.** The *periodogram estimator (sample spectrum)* is given by

$$\hat{S}_x^P(f) = \frac{T}{N} \left| \sum_{n=0}^{N-1} x[n] e^{-2i\pi fTn} \right|^2 \tag{3.1.8}$$

Here, the estimation is based directly on the data (without the intermediate estimation of the autocorrelation), therefrom the term *direct*. In the following we will refer to (3.1.8) simply as *periodogram*.

Note that the periodogram implicitly makes use of windowing: We can imagine a rectangular window $w_N^R[n]$ causing the finite sum in (3.1.8) and alternatively express the periodogram as

$$\begin{aligned}\hat{S}_x^P(f) &= \frac{1}{NT} \left| T \sum_{n=-\infty}^{\infty} w_N^R[n] x[n] e^{-2i\pi fNT} \right|^2 \\ &= \frac{1}{NT} \left| \mathfrak{F}\left\{ w_N^R[n] x[n] \right\} \right|^2, \end{aligned} \tag{3.1.9}$$

i.e. as the normalized squared DTFT of the windowed data. This results in leakage, and therefrom the poor asymptotic properties of the periodogram (Brockwell and Davis 1991).

**Lemma 3.** *Properties of the periodogram.*

1. *The periodogram is biased.*

2. *The periodogram is asymptotically unbiased.*

3. *The periodogram is inconsistent.*

*Proof.* Appendix. □

### 3.1.4.2  Welch power spectral estimation

The Welch power spectral estimation is a widely used non-parametric estimation method and one of the most popular ones in signal processing applications, compare Chapter 5 of Marple (1987).

The *Welch periodogram* constitutes a consequent development of the periodogram estimation approach (3.1.8). As detailed in (3.1.9), the periodogram spectral estimation suffers from strong leakage effects due to implicit usage of rectangular data windows. In order to overcome this limitation, weighted modifications, namely the Daniell periodogram (Daniell 1946) and the Bartlett periodogram (Bartlett 1948, 1950), were brought up. Subsequently Welch (1967) suggested an enhanced estimation approach which unites many of the preceding ideas. We want to mention three important aspects of the Welch periodogram.

**Overlapping segments**  In order to smooth the periodogram, Welch suggested to use *pseudo-ensemble averaging* (see Definition 4 below), just as Bartlett had done. However, in difference to the Bartlett periodogram, segments may overlap each other.

**Windowing**  In order to reduce the leakage effect due the appearance of side lobes, the data of each segment are pre-multiplied with a weighting window. This approach resembles the one of the Daniell periodogram.

**Computational efficiency**  Moreover, Welch provided a computationally efficient procedure for the usage of the FFT in his spectral estimate. Besides its statistical properties, this gain in processing time made the Welch method so popular.

We summarize the Welch spectral estimation method in the following

**Definition 4.** The construction of the *Welch periodogram* is as follows:

1. Divide the data sequence $x[n]$, $n = 1,\ldots,N$ into $P$ (potentially overlapping) segments consisting of $D$ samples with a time shift of $S$ samples from one segment to another. The weighted $p$th segment $x^{(p)}[n]$ then consists of the data samples

$$x^{(p)}[n] = w[n]x[pS+n], \ 0 \leq n \leq D-1.$$

Hereby, each of the $P = \text{floor}\left(\frac{N-D}{S+1}\right)$ segments is weighted with an arbitrary window $w[n]$ of finite energy.

2. A modified sample spectrum is calculated for each segment $x^{(p)}[n]$,

$$\tilde{S}_{xx}^{(p)}(f) = \frac{T}{UD} \left| \sum_{n=0}^{D-1} x^{(p)}[n] e^{-2i\pi fTn} \right|^2 \quad \forall p, 0 < p \leq P-1.$$

The normalization constant $U$ is the power of the window $w[n]$, i.e.

$$U = \frac{T}{D} \sum_{n=0}^{D-1} w[n]^2.$$

It ensures that the Welch periodogram is asymptotically unbiased, compare Lemma 4.

3. The Welch averaged periodogram (at each frequency point $f$) is given by

$$\hat{S}_x^W(f) = \frac{1}{P} \sum_{p=0}^{P-1} \tilde{S}_x^{(p)}(f). \tag{3.1.10}$$

We refer to Welch (1967) and Barbé et al. (2010) for a detailed statistical analysis and summarize important properties of the Welch periodogram:

**Lemma 4.** *Properties of the Welch periodogram.*

1. *The Welch periodogram is biased.*

2. The Welch periodogram is asymptotically unbiased.

3. The variance of the Welch periodogram is inversely proportional to the number of segments.

*Proof.* Appendix. □

We want to end this section with two remarks regarding estimation quality.

First, as mentioned in Subsection 3.1.1, we have to choose a trade-off between the frequency resolution and the variance: The more segments we allow, the higher the spectral resolution, but the »noisier«, i.e. the more fluctuating, the Welch estimator becomes.

Second, estimation quality also depends on the data window. For instance, Welch (1967) suggested to employ the Hann window. However, due to the excellent properties of the Hamming window in reducing the leakage effect (relative peak side-lobe amplitude of -41 dB, see Table 3.1.1), we use the Hamming window in Welch spectral estimation throughout this study.

## 3.2 Autoregressive modeling

> The problem of determining the period and the disturbances, in the case of the sunspot numbers, was attacked in the first instance [...] by finding the best (least square) linear equation relating $u_x + u_{x-2}$ to $u_{x-1}$, this giving the form of difference equation required for a simple harmonic function.
>
> — Yule (1927): *On a Method of Investigating Periodicities in Disturbed Series, with Special Reference to Wolfer's Sunspot Numbers*, foundations of autoregressive modeling

A common approach in neuroscience literature (Franaszczuk et al. 1985, Blinowska and Kaminski 2006) is to model EEG recordings as multivariate autoregressive processes. This popularity results from the fact that AR spectra show sharp peaks, constitute high-resolution spectral estimates (compare Section 7.4 of Marple (1987)) and are obtained with little computational complexity.

Assume that the signal $\mathbf{x}[n]$ of interest is stationary with mean zero. We consider the following class of models:

**Definition 5.** An *AR model of order p* is given by

$$\mathbf{x}[n] = \sum_{s=1}^{p} \mathbf{A}[s]\mathbf{x}[n-s] + \varepsilon[n], \quad (3.2.1)$$

where $\varepsilon[n]$ is white noise with regular covariance matrix $\Sigma_\varepsilon$ and the AR coefficients $\mathbf{A}[s]$ are matrices of dimension $K \times K$ for each lag $s$.[6] We will refer to such a model as *regular AR(p) model*.

Using $z$ to denote both a complex variable and the backward-shift-operator[7], we introduce the polynomial matrix $\mathbf{A}(z) = \mathbf{I}_{K \times K} - \tilde{\mathbf{A}}(z)$ with the complex power series $\tilde{\mathbf{A}}(z) = \sum_{s=1}^{p} \mathbf{A}[s] z^p$. This allows to rewrite the AR model (3.2.1) in compact notation.

**Definition 6.** The AR(p) model (3.2.1) can be written in polynomial form as

$$\mathbf{A}(z)\mathbf{x}[n] = \varepsilon[n], \quad (3.2.2)$$

where we assume the *stability condition* to hold:

$$\det \mathbf{A}(z) \neq 0, |z| \leq 1. \quad (3.2.3)$$

As the stability condition (3.2.2) assures the invertibility of the polynomial matrix $\mathbf{A}(z)$, we obtain

$$\mathbf{x}[n] = \mathbf{A}(z)^{-1} \varepsilon[n] = \mathbf{H}(z) \varepsilon[n]$$

as solution of (3.2.1). Evaluation of $\mathbf{A}(z)$ at $z = \exp(2i\pi f)$ yields the complex-valued model coefficients $\mathbf{A}(f)$ and the *transfer function* $\mathbf{H}(f) = \mathbf{A}(f)^{-1}$ in the frequency domain (Oppenheim and Schafer 1989).

---

[6]Note that in (3.2.1) the coefficient matrix is the identity. Thus, any further specification of $\Sigma_\varepsilon$ (for instance, assuming diagonal form) would restrict the process.

[7]We draw attention to the fact that in electrical engineering $z$ is often inversely defined as forward-shift-operator.

Note that in applications one would choose the following way to calculate $\mathbf{A}(f)$: Once the AR model has been identified (by solving the Yule-Walker equations), $\mathbf{A}(f)$ and $\mathbf{H}(f)$ are obtained by Fourier transformation of the model coefficients, i.e.

$$\mathbf{A}(f) = \mathbf{I}_{K \times K} - \tilde{\mathbf{A}}(f) = \mathbf{I}_{K \times K} - \mathfrak{F}\mathbf{A}[s]. \qquad (3.2.4)$$

Here, we employ the discrete-time Fourier transform of the AR model coefficients as a special case of the z-transform in (3.2.2). We will need representation (3.2.4) in the context of dependency measures, see Section 3.4.

The (matrix-valued) power spectral density directly follows:

$$\mathbf{S_x}(f) = \mathbf{H}(f)\,\mathbf{S}_{\varepsilon}\mathbf{H}(f)^H = \frac{1}{2\pi}\mathbf{H}(f)\Sigma_{\varepsilon}\mathbf{H}(f)^H, \qquad (3.2.5)$$

where the diagonal elements represent the auto-spectra and the off-diagonal elements the complex-valued cross-spectra.

## 3.3 Granger causality

> We say that $Y_t$ is causing $X_t$ if we are able to better predict $X_t$ using all available information than if the information apart from $Y_t$ had been used.
>
> — *Investigating causal relations by econometric models and cross-spectral methods* by Granger (1969), Nobel Memorial Prize in Economic Sciences in 2003

Throughout the last decades there have been long and thorough discussions how causality can be formalized mathematically. A brief summary of the concept of causality can be found in Pearl (2000). In this study we will limit ourselves to Granger causality, as introduced in Granger (1969), based on a suggestion by Wiener (1956).

Note that we only give a short overview of the fundamental ideas of Granger causality in this section. In case the reader is interested in a more theoretical introduction, we refer to the doctoral thesis of Flamm (2012).

### 3.3.1 Conditional Granger Causality

According to the definition of Granger (1969), we say a signal $x_1$ is causing another signal $x_2$, denoted by $x_1 \to x_2$, if knowledge of $x_1$'s past significantly improves the prediction of $x_2$. Therefore, predicting $x_2$ from its own past and the one of $x_1$ leads to a decrease of the prediction error in this case, compared to a prediction of $x_2$ from its own past.

For our purposes we consider the multivariate extension of this bivariate concept according to Eichler (2007), which is often referred to as *conditional Granger causality*.

**Definition 7** (Conditional Granger Causality in AR-framework). Let $\mathbf{x}[n]$ be AR-modeled according to (3.2.1).

- We say that $x_i$ is Granger-non-causal for $x_j$ with respect to $\mathbf{x}$, denoted by $x_i \nrightarrow x_j | \mathbf{x}$, if $\mathbf{A}_{ji}(z) = 0$, i.e. $\mathbf{A}_{ji}[s] = 0 \, \forall s$.

- We say that $x_i$ is Granger-causal for $x_j$ with respect to $\mathbf{x}$, denoted by $x_i \to x_j | \mathbf{x}$, if $\mathbf{A}_{ji}(z) \neq 0$, i.e. $\mathbf{A}_{ji}[s] \neq 0$ for at least one lag $s$.

This means that in case of Granger causality, i.e. $x_i \to x_j | \mathbf{x}$, $x_j$ influences $x_i$ in the AR representation (3.2.1). Thus, knowledge of $x_i$'s past improves the prediction of $x_j$, in analogy to the original bivariate definition of Granger (1969). As we only consider off-diagonal elements ($i \neq j$) for this kind of analysis, we can regard $\mathbf{A}(z)$ instead of $\tilde{\mathbf{A}}(z)$.

As in applications the AR coefficients will hardly ever be zero, one has to test statistically whether they differ from zero significantly. For this purpose, Eichler (2006a) derives a $\chi^2$-test, and Seth (2010) implements an $F$-test in his Matlab toolbox for Granger causality analysis.

Note that conditional Granger causality is a meaningful extension to the bivariate case, as in a bivariate AR model the definition of conditional Granger causality simply reduces to the bivariate definition of Granger (1969). Again, the AR coefficients indicate Granger causality or Granger non-causality.

### 3.3.2 Granger Causality in the frequency domain

The notion of Granger Causality is closely related to the time domain, as it is based on prediction. Here a concept of Granger Causality in the frequency domain shall be introduced.

Based on the work of Geweke (1982, 1984), who led the foundations for a transfer of the concept of Granger causality into the frequency domain, Baccala and Sameshima (2001) were interested in a »frequency-domain picture« of this notion. Their motivation originates from neuroscience: Depending on the patient, different frequency bands reveal different neurological phenomena. It might therefore be interesting to regard »Granger causality at a certain frequency« (Baccala and Sameshima 2001).[8]

According to Subsection 3.3.1, Granger causality can be determined in an AR framework by looking at the coefficients $\mathbf{A}[s]$ or the polynomial coefficient matrix $\mathbf{A}(z)$. Thus, according to (3.2.4), $\mathbf{A}(f) = \mathbf{I}_{K \times K} - \mathfrak{F}\mathbf{A}[s]$ indicates Granger causality for off-diagonal elements $A_{ji}(f), i \neq j$ in the frequency domain.[9]

If $A_{ji}(f) = 0 \,\forall f$ we conclude that $x_i[n]$ does not Granger cause $x_j[n]$; if $A_{ji}(f) = 0$ for a certain $f$, we follow the approach of Baccala and Sameshima (2001) and say that »$x_i[n]$ does not Granger cause $x_j[n]$ at frequency $f$«. In order to derive this statement for a frequency band $[f_1, f_2]$, we simply consider $\int_{f_1}^{f_2} |A_{ji}(f)| \, df$.

## 3.4 Dependency measures

> In discussions of the relations between time series, concepts of dependence and feedback are frequently invoked.
>
> — Geweke (1982): *Measurement of Linear Dependency and Feedback Between Multiple Time Series,* basis of modern dependency measures in neuroscience

---

[8] The idea of a frequency-domain description of Granger causality is exploited mathematically in a recent review by Ding et al. (2006). In particular, Chen et al. (2006) derive a non-parametric estimation approach to Granger causality in the frequency domain.

[9] Note that we look at $\mathbf{A}(z)$ evaluated at $z = \exp(2i\pi f)$, compare Section 3.2.

The goal of applying dependency measures to the autoregressive framework (3.2.1) is to gain insights into the inner causal structure of the multivariate signal, i.e. to derive statements which components influence each other. This is particularly important in neuroscience, as it allows to conclude which channels of EEG recordings reveal dependencies, thus indicate couplings between different brain regions. The results of this analysis can then be visualized in a graph where the vertices represent the channels and the edges the dependencies, compare Edwards (2000) for an introduction to graph theory. We are concerned with two major aspects in this type of graphical analysis:

**Directed vs undirected:** A dependency can have a notion of directedness or not. In case a directed measure indicates an influence from e.g. $x_1$ to $x_2$, we say $x_1 \to x_2$, and we draw an arrow in the graph from node $x_1$ to node $x_2$. In case an undirected measure indicates an influence, we say $x_1$ and $x_2$ are coupled and simply connect vertices $x_1$ and $x_2$ by an (undirected) edge. In any case, the absence of coupling results in separated, i.e. not connected, vertices.

**Direct vs indirect:** A direct dependency measure only indicates couplings which affect two neighboring nodes, but which are not mediated via a third node. Assume three nodes $x_1$, $x_2$ and $x_3$. An information flow from $x_1$ via $x_3$ to $x_2$ would be suppressed in this case. In the opposite case of indirect dependency measures both direct and indirect couplings are considered.

In this subsection we discuss prominent frequency-domain dependency measures regarding the aforementioned properties, based on the multivariate autoregressive modeling framework (3.2.1). Thus, we will consider a zero-mean, stationary, $K$-dimensional signal $\mathbf{x}[n]$ throughout this section. We assume that we have identified the AR model (3.2.1) and know the matrices $\mathbf{A}(f)$, $\mathbf{H}(f)$ and $\mathbf{S}(f)$, compare Section 3.2. Note that it is of primary importance that this dependency analysis is performed in a fully multivariate way, i.e. that a multivariate AR model is identified at once, as subsequent bivariate analysis steps might lead to erroneous dependency results (Kus et al. 2004).

We refer to Flamm et al. (2012) for further theoretical considerations regarding dependency analysis in a multivariate autoregressive modeling framework.

As we stick to a linear framework in this study, we do not consider non-linear methodology like, for example, non-linear causality analysis. Note, however, that the latter has often been applied successfully to neurophysiological problems, as recently reported by e.g. Chavez et al. (2003), Chen et al. (2004) and Gourévitch et al. (2006). A good overview of non-linear dependence measures is provided in Pereda et al. (2005).

### 3.4.1 Ordinary Coherence

The (ordinary) coherence is, in a way, the most natural coupling measure based on spectral properties. It quantifies the normalized dependency of two signal components at a certain frequency, compare Brockwell and Davis (1991).

**Definition 8** (Ordinary Coherence). The Ordinary Coherence (OC) of $x_i$ and $x_j$ is given by

$$C_{i,j}^2(f) \triangleq \frac{|S_{ij}(f)|^2}{S_{ii}(f)\,S_{jj}(f)}, \tag{3.4.1}$$

where $S_{ij}(f)$ is the $(i,j)$-entry of the spectral matrix $\mathbf{S}(f)$.

As can be seen immediately from the Cauchy-Schwarz-inequality, the Ordinary Coherence (3.4.1) is bounded by 0 and 1, where a value of 0 signifies absence of coupling and a value of 1 perfect linear dependence at frequency $f$. Due to its symmetric definition, it obviously does not involve any notion of directedness; neither is it able to distinguish between direct and indirect influences.

### 3.4.2 Partial Coherence

In order to obtain a direct dependency measure, the Ordinary Coherence is extended to the Partial Coherence (PC). Its definition is in analogy to the one of the Ordinary Coherence, but makes use of the so-called *partialized spectrum*. The approach of Dahlhaus (2000) and Brillinger (2001) to derive the latter is as follows:

Let $x_i[n]$ and $x_j[n]$ be two channels of the multivariate signal $\mathbf{x}[n]$. The other components are denoted by $\mathbf{y}_{i,j}[n] = \{x_k[n], k \neq i, j\}$. In order to shorten the notation we will sloppily write $\mathbf{y}[n]$, though dependent on $i$ and $j$.

Using this notation, the following steps are performed:

1. First, consider the residuals $\varepsilon_i[n]$ and $\varepsilon_j[n]$ which are the respective components, with the influence of all other components removed

$$\begin{cases} \varepsilon_i[n] \triangleq x_i[n] - \mathbf{d}_i^T * \mathbf{y} = x_i[n] - \sum_{\tau=-\infty}^{\infty} \mathbf{d}_i^T[n-\tau]\mathbf{y}[\tau] \\ \varepsilon_j[n] \triangleq x_j[n] - \mathbf{d}_j^T * \mathbf{y} = x_j[n] - \sum_{\tau=-\infty}^{\infty} \mathbf{d}_j^T[n-\tau]\mathbf{y}[\tau] \end{cases}, \quad (3.4.2)$$

where $\mathbf{d}_i$ and $\mathbf{d}_j$ symbolize optimal filters.

2. The partial covariance is then calculated as $\text{cov}(\varepsilon_i[n], \varepsilon_j[n])$, and the partial cross-spectrum is its Fourier transform

$$S_{i,j|\mathbf{y}}(f) = \mathfrak{F}\{\text{cov}(\varepsilon_i[n+s], \varepsilon_j[n])\}. \quad (3.4.3)$$

The definition of the *Partial Coherence* immediately follows:

**Definition 9** (Partial Coherence)**.** The Partial Coherence (PC) of $x_i$ and $x_j$ is given by normalization of the partial spectrum (3.4.3)

$$R^2_{i,j|\mathbf{y}}(f) \triangleq \frac{|S_{i,j|\mathbf{y}}(f)|^2}{S_{i,i|\mathbf{y}}(f) S_{j,j|\mathbf{y}}(f)}. \quad (3.4.4)$$

Like the Ordinary Coherence, the Partial Coherence is bounded by 0 and 1, and a value of 0 symbolizes absence of coupling. Confidence intervals for the Partial Coherence are given in e.g. Winterhalder et al. (2005). Similar to the Ordinary Coherence, the Partial Coherence does not involve any notion of directedness due to its symmetric definition.

Note that the Partial Coherence (3.4.4) can be obtained in a quicker way, see Dahlhaus (2000): Instead of performing the partialization of channels $x_i$ and $x_j$, one can simply invert the ordinary spectrum and re-normalize it appropriately.

The above methodology is purely non-parametric, thus does not fit into the AR model framework (3.2.1) However, an explicit formula involving $\mathbf{A}(f)$, $\mathbf{H}(f)$ and $\mathbf{S}(f)$ would be desirable, as the goal is to consider dependency measures within the framework of autoregressive modeling.

Korzeniewska et al. (2003) provide a simple formula for the calculation of the Partial Coherence (3.4.4) using $\mathbf{S}(f)$:

**Lemma 5** (Computational formula for the Partial Coherence). *The partial coherence (PC) in the AR modeling framework (3.2.1) can be computed as*

$$\chi_{i,j}^2(f) = \frac{|M_{i,j}(f)|^2}{M_{i,i}(f) M_{j,j}(f)}, \tag{3.4.5}$$

*where $M_{i,j}(f)$ denotes the minor of $\mathbf{S}(f)$ which is obtained by removing row i and column j from $\mathbf{S}(f)$.*

*Proof.* Appendix. □

Note, however, that in practical applications the results of formula (3.4.5) will slightly differ from Dahlhaus' approach (3.4.4). This is due to the fact that a measured biological signal is not the exact realization of an autoregressive process, although an AR model might be fitted well to the data. In other words, Dahlhaus' approach (3.4.4) yields the Partial Coherence of the ECoG signal whereas (3.4.5) is the Partial Coherence of the process specified by the AR model.

### 3.4.3 Directed Transfer Function

As already mentioned, Ordinary Coherence and Partial Coherence do not involve any notion of directedness. In order to overcome this limitation, Kaminski and Blinowska (1991) proposed a dependency measure termed *Directed Transfer Function (DTF)*. Although its exact mathematical interpretation remains unclear (see below), it is often used in neuroscience literature (Blinowska 2011).

#### 3.4.3.1 DTF

The Directed Transfer Function is based on the transfer function $\mathbf{H}(f)$, as its name indicates:

**Definition 10** (Directed Transfer Function). The Directed Transfer Function (DTF) is defined as

$$\gamma_{i,j}^2(f) \triangleq \frac{|H_{ij}(f)|^2}{\sum_{m=1}^{K}|H_{im}(f)|^2}. \tag{3.4.6}$$

As the normalization in (3.4.6) is performed with respect to all source channels $x_m$, $m = 1, \ldots, K$, DTF can be interpreted as the ratio of the inflow from channel $x_j$ to $x_i$ normalized to all inflows to $x_i$. The inflow from $x_j$ to $x_i$ comprises both direct and indirect information flows, compare Eichler (2006b): The numerator of DTF represents the total inflow, as its expansion as geometric series yields

$$\begin{aligned}\mathbf{H}(f) &= \mathbf{A}(f)^{-1} = \left(\mathbf{I}_{K \times K} - \tilde{\mathbf{A}}(f)\right)^{-1} \\ &= \sum_{m=0}^{p} \mathbf{A}(f)^m = \mathbf{I}_{K \times K} + \tilde{\mathbf{A}}(f) + \tilde{\mathbf{A}}(f)^2 + \ldots\end{aligned}$$

For off-diagonal elements, $i \neq j$, we obtain

$$H_{ij}(f) = \tilde{A}_{ij}(f) + \sum_{w}\tilde{A}_{iw}(f)\tilde{A}_{wj}(f) + \ldots, \tag{3.4.7}$$

where $A_{ij}(f)$ represents the direct inflow from $x_j$ to $x_i$ and the subsequent terms all other possible indirect information flows.

DTF is bounded by 0 and 1, as

$$\sum_{n=1}^{K} \gamma_{i,n}^2(f) = \sum_{n=1}^{K} \frac{|H_{in}(f)|^2}{\sum_{m=1}^{K} |H_{im}(f)|^2} = 1.$$

Low values indicate weak influence from $x_j$ to $x_i$, values near 1 high influence; a value of 0 symbolizes absence of coupling. In order to test statistically whether DTF differs from zero, confidence intervals have to be considered. However, exact ones are not available. Kaminski et al. (2001) propose surrogate data methods for constructing confidence intervals based on numerical simulations (bootstrapping), and Eichler (2006b) derives an asymptotic confidence interval.

Obviously, DTF (3.4.15) is not a symmetric measure. The authors of DTF even claim that it indicates directed couplings: As $\mathbf{H}(f)$ is not symmetric, it contains information on the information transmission from $x_j$ to $x_i$. From this fact they conclude »that off-diagonal elements of $\mathbf{H}(f)$ can be a measure of the directional flow that we are seeking« (Kaminski et al. 2001).

In a special case DTF is linked to the notion of Granger causality (Kaminski et al. 2001):

**Lemma 6** (DTF and bivariate Granger Causality). *The DTF (3.4.15) indicates Granger causality in the bivariate case: If $\gamma_{ij}^2(f) = 0 \ \forall f$, then $x_j \not\to x_i$.*

*Proof.* Appendix. □

Note that this still holds for multivariate signals $\mathbf{x}[n]$ partitioned into two component signals $\mathbf{x}_I[n]$ of dimension $M$ and $\mathbf{x}_J[n]$ of dimension $N$ with $M + N = K$. DTF then indicates Lütkepohl (2007)'s extension of bivariate Granger causality, as can be seen from a direct generalization of the proof of theorem 6.

However, this equivalence is not valid in the general multivariate case. Although Kaminski et al. (2001) claim that the situation »remains unclear«, counter-examples of Eichler (2006b) prove the opposite, compare the discussion of DTF below.

For reasons of completeness note that DTF (3.4.6) is a special case of a measure termed *Directed Coherence*. The latter is defined as

$$\tilde{\gamma}_{i,j}^2(f) \triangleq \frac{\sigma_{jj}^2 |H_{ij}(f)|^2}{\sum_{m=1}^{K} \sigma_{jm}^2 |H_{im}(f)|^2}, \qquad (3.4.8)$$

which gives DTF for $\Sigma = \mathbf{I}_{K \times K}$. For further details we refer to Baccala et al. (1998).

### 3.4.3.2 Extensions

DTF involves a notion of directedness, but cannot distinguish between direct and indirect dependencies, see derivation (3.4.7). Korzeniewska et al. (2003) therefore constructed a measure combining these two desired properties. This construction is performed in an intuitive way in two steps.

First, consider the following measure.

**Definition 11** (Full frequency Directed Transfer Function). The full frequency Directed Transfer Function (ffDTF) is defined as

$$\eta_{i,j}^2(f) \triangleq \frac{|H_{ij}(f)|^2}{\int_f \sum_{m=1}^{K} |H_{im}(f)|^2}. \qquad (3.4.9)$$

Second, the *direct Directed Transfer Function (dDTF)* is introduced as the product of the Partial Coherence (3.4.5) and the full frequency Directed Transfer Function (3.4.9):

**Definition 12** (Direct Directed Transfer Function). The direct Directed Transfer Function (dDTF) is given by

$$\text{dDTF} \triangleq \text{PC} \cdot \text{ffDTF}. \qquad (3.4.10)$$

This construction has the advantage that the Partial Coherence sorts out indirect couplings and the full frequency Directed Transfer Function contributes the notion of directedness.

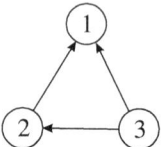

**Figure 3.4.1:** *DTF and Granger causality.* Dependencies of variables in the AR system (3.4.11) given by Eichler (2006b) as a counter-example to the claimed link between DTF and conditional Granger causality.

Moreover, Ginter et al. (2001) extended DTF to its time-variant form termed *short-term DTF (sDTF)*, which has successfully been applied to the analysis of cognitive tasks (Blinowska et al. 2010). However, it is not applicable in the context of epileptic seizure propagation, as its calculation requires repetitions of the same experimental setup (e.g. repeated execution of the same cognitive task).

### 3.4.3.3 Discussion

In the following we briefly discuss three aspects of DTF which merit, in our opinion, some attention.

**Causality:** Although bivariate DTF indicates Granger causality, this is not true for conditional Granger causality. Eichler (2006b) gives a simple counterexample: Consider the multivariate autoregressive system

$$\begin{cases} x_1[n] &= \alpha x_2[n-1] + \beta x_3[n-2] + \varepsilon_1[n] \\ x_2[n] &= \gamma x_3[n-1] + \varepsilon_2[n] \\ x_3[n] &= \varepsilon_3[n] \end{cases}, \quad (3.4.11)$$

whose dependency paths are represented in Fig. 3.4.1.

A short calculation shows that the numerator of DTF $\gamma_{1,3}^2(f)$ is given by $(\beta + \alpha\gamma)^2$. We distinguish two cases:

- $\beta = 0$
  As the AR coefficient $A_{13} = 0$, $x_3$ does not Granger-cause $x_1$. However, DTF takes a value different from zero, $\gamma_{1,3}^2(f) = \alpha^2\gamma^2$,

i.e. indicates Granger causality.

This contradiction is caused by the indirect influence of $x_3$ via $x_2$ to $x_1$, also compare (3.4.7).

- $\beta = -\alpha\gamma \neq 0$

  In this case, the situation is the exact opposite: DTF equals zero, but $x_3$ does Granger-cause $x_1$.

**Transfer function:** Second, consider the following system theoretic interpretation: Based on the autoregressive model (3.2.1), $\gamma_{i,j}^2(f)$ describes the information transfer from the innovations of channel $x_j$, i.e. $\varepsilon_j$, to channel $x_i$. However, we are interested in couplings between channels $x_i$ and $x_j$, not between $x_i$ and $\varepsilon_j$.

**Normalization:** Finally note that it is not obvious why DTF is normalized to all signal sources. While Kaminski and Blinowska (1991) state that they search for a normalization which puts emphasis on the signal structures sending the signal, one could also normalize to any other structure, e.g. target channels.

### 3.4.4 Partial Directed Coherence

Another prominent coupling indicator in neuroscience is the *Partial Directed Coherence (PDC)*, introduced by Baccala and Sameshima (2001) for providing a »frequency-domain picture« of Granger causality. As its name already indicates, it was constructed in order to combine the two properties we are seeking: a notion of directedness and suppression of indirect information flow.

#### 3.4.4.1 PDC

We follow Baccala and Sameshima (2001)'s construction of PDC step by step in order to understand to which extent its mathematical properties are rigorous and to which heuristic.

First, consider an alternative expression of the Partial Coherence (3.4.4).

**Definition 13** (Partial Coherence Function). The Partial Coherence Function is given by

$$\kappa_{i,j}(f) \triangleq \frac{\mathbf{a}_{\cdot i}^H(f) \Sigma^{-1} \mathbf{a}_{\cdot j}(f)}{\sqrt{\mathbf{a}_{\cdot i}^H(f) \Sigma^{-1} \mathbf{a}_{\cdot i}(f)} \sqrt{\mathbf{a}_{\cdot j}^H(f) \Sigma^{-1} \mathbf{a}_{\cdot j}(f)}}, \qquad (3.4.12)$$

where $\mathbf{a}_{\cdot i}(f)$ denotes the ith column of the matrix $\mathbf{A}(f)$.

**Lemma 7** (Partial Coherence Function and PC). *The Partial Coherence Function $\kappa_{i,j}(f)$ is the Partial Coherence (3.4.4).*

*Proof.* Appendix. □

Second, consider a factorization of the Partial Coherence:

The Partial Coherence Function is a partialized, but undirected measure. By factorizing $\kappa_{i,j}(f)$, the authors claim to overcome this limitation. For this purpose they define the *Partial Directed Coherence Factor (PDCF)*.

**Definition 14** (Partial Directed Coherence Factor). The Partial Directed Coherence Factor (PDCF) is given by

$$\pi_{i,j}(f) \triangleq \frac{A_{ij}(f)}{\sqrt{\mathbf{a}_{\cdot j}^H(f) \Sigma^{-1} \mathbf{a}_{\cdot j}(f)}}, \qquad (3.4.13)$$

where $\mathbf{a}_{\cdot i}(f)$ again denotes the ith column of the matrix $\mathbf{A}(f)$.

This definition yields the desired factorization:

**Lemma 8** (Factorization of the Partical Coherence Function). *The Partial Coherence Function can be factored as*

$$\kappa_{i,j}(f) = \pi_{\cdot i}^H(f) \Sigma^{-1} \pi_{\cdot j}(f), \qquad (3.4.14)$$

where $\pi_{\cdot i}(f)$ denotes the ith column of the matrix $\Pi = (\pi_{ij})_{i=1,\ldots,K,\, j=1,\ldots,K}$.

*Proof.* Appendix. □

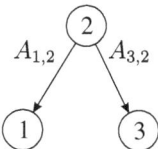

**Figure 3.4.2: *Normalization of the PDC*.** Exemplary 3-dimensional system with source channel $x_2$. $\pi_{1,2}^2(f)$ and $\pi_{3,2}^2(f)$ are both normalized with respect to $|A_{1,2}(f)^2| + |A_{2,2}(f)^2| + |A_{3,2}(f)^2|$, thus to all outflows from $x_2$.

Note that obviously the Partial Directed Coherence Factor (3.4.13) is an asymmetric measure. Baccala and Sameshima (2001) even claim that this factorization into two components introduces a notion of directedness.

The last step in the construction of PDC is of heuristic nature: In (3.4.13), $\Sigma$ affects the denominator of the Partial Directed Coherence Factor. Unfortunately, as $\Sigma$ describes the instantaneous coupling between the signal components, (3.4.13) would indicate mixed effects of both Granger causality and instantaneous coupling.

In order to overcome this limitation, the authors define the *Partial Directed Coherence* as follows:

**Definition 15** (Partial Directed Coherence). The Partial Directed Coherence (PDC) is given by

$$\pi_{i,j}^2(f) \triangleq \frac{|A_{ij}(f)|^2}{\sum_{n=1}^{K} |A_{nj}(f)|^2}. \qquad (3.4.15)$$

which is identical to the Partial Directed Coherence Factor (3.4.13) in the case of $\Sigma = \mathbf{I}_{K \times K}$.

As the normalization in (3.4.15) is performed with respect to all target channels, PDC can be interpreted as the ratio of the outflow from channel $x_j$ to $x_i$ normalized to all outflows from $x_j$. Emphasis is therefore put on the neural structure sending the signal, as shown in Fig. 3.4.2. Note that this normalization is exactly inverse to the one of DTF.

Similar to DTF, PDC is bounded by 0 and 1, as

$$\sum_{m=1}^{K} \pi_{mj}^2(f) = \sum_{m=1}^{K} \frac{|A_{in}(f)|^2}{\sum_{n=1}^{K} |A_{nj}(f)|^2} = 1.$$

Low values indicate little influence from $x_j$ onto $x_i$, and values near 1 high influence. A value of 0 symbolizes absence of coupling. Again, one has to test statistically whether PDC differs from zero significantly. Besides an asymptotic (frequency-dependent) confidence interval derived by Schelter et al. (2005)[10], Takahashi et al. (2007) provide an exact one.

Unlike DTF, PDC is linked to conditional Granger causality (and therefore to Granger causality in the bivariate case as well):

**Lemma 9** (PDC and Conditional Granger Causality). *PDC (3.4.15) indicates Conditional Granger Causality in the multivariate AR model (3.2.1):*

$$\pi_{i,j}^2(f) = 0 \,\forall f \iff x_j[n] \nrightarrow x_i[n].$$

*Proof.* Appendix. □

Therefore, PDC provides a description of Granger causality in the frequency domain, what (Baccala and Sameshima 2001) call »frequency-domain picture«, compare Subsection 3.3.2.

Finally, for the sake of completeness note that PDC is well-defined (Schelter et al. 2005).

**Lemma 10** (Well-definedness of the PDC). *If the stability condition (3.2.2) is satisfied for the autoregressive model (3.2.1), the denominator of PDC does not equal zero:*

$$\sum_{n=1}^{K} |A_{nj}(f)|^2 \neq 0.$$

*Thus, PDC is well-defined.*

*Proof.* Appendix. □

---

[10] The disadvantage of this confidence interval lies in the fact that it is only valid point-wise: We therefore expect exceedings of the confidence level even in case of non-significance, compare Schelter et al. (2005). In order to overcome this limitation Sommerlade, Eichler, Jachan, Henschel, Timmer and Schelter (2009) proposed a smoothed version of PDC.

### 3.4.4.2 Extensions

Due to the enormous interest in PDC in the neuroscience community various extensions of its initial definition (3.4.15) have come up in the last years.

First, Baccala et al. (2007) defined *generalized PDC (gPDC)* in order to overcome the limitation of PDC that it is not scale-variant, i.e. depends on the scaling of the individual signal components.

**Definition 16** (Generalized Partial Directed Coherence). gPDC is given as

$$gPDC_{i,j} \triangleq \frac{\frac{1}{\sigma_i^2}\left|A_{ij}(f)\right|^2}{\sum_{n=1}^{K}\frac{1}{\sigma_n^2}\left|A_{nj}(f)\right|^2},$$

where $\sigma_i^2$ the entry $\Sigma_{ii}$ of the error covariance matrix. Note that this renormalization preserves the boundedness by 1.

Second, Takahashi et al. (2010) derived an information theoretic form of PDC establishing a link to mutual information. Finally, PDC has recently been extended to time-variant forms, compare Sommerlade, Henschel, Wohlmuth, Jachan, Amtage, Hellwig, Lücking, Timmer and Schelter (2009) and Omidvarnia et al. (2012).

### 3.4.4.3 Discussion

As for DTF, we want to end this subsection on PDC with a short discussion on its construction and properties.

**Directedness:** First, Baccala and Sameshima (2001) claim that factorization of the Partial Coherence Function (3.4.12) into Partial Directed Coherence Factors (3.4.13) yields a directed measure. Obviously the first is symmetric, i.e. undirected, and the latter is asymmetric. While asymmetry does not necessarily imply causality, we do know that PDC involves a notion of directedness due to its link to conditional Granger causality.

**Omission of $\Sigma$:** Second, PDC is obtained from the Partial Directed Coherence Factor (3.4.13) by omission of $\Sigma$ in the denominator. In case of $\Sigma = \mathbf{I}_{K \times K}$ the derivation of PDC is accurate.

**Normalization:** Finally, the normalization with respect to all target channels is one possible way among many. One could, for instance, also choose all source structures like DTF does.

In particular, the specific kind of normalization with respect to all target channels poses problems, compare Schelter et al. (2009): When considering the coupling between two channels, a third channel not even involved in the consideration affects the denominator of the PDC and thus changes the coupling quantification. Compare Fig. 3.4.2 for an illustration. Obviously, analog limitations apply when normalizing with respect to all source channels as in case of DTF (Graef et al. 2009).

### 3.4.5 Granger Causality Index

The last coupling indicator we want to present is directly related to Granger causality, the *measure of conditional linear dependence* (Geweke 1984).[11] In neuroscience literature this measure is sometimes referred to as *Granger causality index (GCI)*, see e.g. Winterhalder et al. (2005).

For the motivation of this coupling indicator we have to recall the basic idea of Granger causality. So far, we have only considered the characterization of conditional Granger causality via the AR coefficients, compare subsection 3.3.1. Here we exploit Granger (1969)'s basic definition saying that knowledge of one signal's past significantly improves the prediction of another one. We consider an extension of to the multivariate case (Eichler 2007): We denote the prediction error of $x_j$, given all other components, by $\sigma^2(x_j|\mathbf{x})$ and the prediction error of $x_j$, given all other components except

---

[11] This measure is an extension of the *measure of linear dependence* for the bivariate case, as introduced in Geweke (1982).

of $x_i$, by $\sigma^2(x_j|\mathbf{x}\setminus\{x_i\})$.[12] In case $x_i$ is Granger-causal for $x_j$, the first prediction error is smaller than the second; in case of non-causality they take the same value.

To quantify this aspect, we simply construct the log ratio of these two terms, compare Geweke (1984),

$$\mathscr{F}_{x_i \to x_j | \mathbf{x}} = \ln \frac{\sigma^2(x_j|\mathbf{x}\setminus\{x_i\})}{\sigma^2(x_j|\mathbf{x})}. \qquad (3.4.16)$$

In case of non-causality it values zero (both regressions yield the same prediction error), otherwise it measures the coupling strength.

We refer to Geweke (1984) for a statistical test. However, in this study we will use the *GCCA toolbox* described in Seth (2010), which makes use of measure (3.4.16) for testing the null hypothesis $\mathscr{H}_0 : A_{ji}[s] = 0\ \forall s$ of conditional Granger non-causality (compare Subsection 3.3.1).

## 3.5 Signal detection

> Every test of a statistical hypothesis [...], consists in a rule of rejecting the hypothesis when a specified character, $x$, of the sample lies within certain critical limits, and accepting it or remaining in doubt in all other cases.
>
> — *On the Problem of the Most Efficient Tests of Statistical Hypotheses* by Neyman and Pearson (1933), statisticians and founding figures of detection theory

The aim of signal detection is to decide whether a specific information is contained in a measured signal, e.g. in radar applications to decide whether the echo of an aircraft is present in the measured signal or not.

We assume that the reader is familiar with basic signal processing techniques and statistical concepts. We refer to Kay (1998) for an introduction to signal detection and to Scharf (1991) for a more theoretical approach to the topic.

---

[12] Compare the concept of partialization detailed in Subsection 3.4.2.

## 3.5.1 Preliminary definitions

In the following we consider a (continuous) $N$-dimensional random variable $\mathbf{x}$, which is described by the probability density function $p_\theta(\mathbf{x})$. The observations $x_n, n = 1, \ldots, N \sim p_\theta(\mathbf{x})$ form the univariate signal $x[n]$ of length $N$ under consideration. The goal is to decide from which (finite) parameter set $\Theta_i$ the parameter $\theta$ was drawn, given the observation space $\mathscr{X} = \{(x_1, \ldots, x_N)\}$.

For a formalization we consider the following definitions:

**Definition 17** (Distinction via number of hypothesis). We distinguish two partitions of the parameter space:

- In case of $\Theta = \Theta_0 \cup \Theta_1$ with $\Theta_0 \cap \Theta_1 = \{\}$, we speak about a *binary hypothesis test*: $\mathscr{H}_0 : \theta \in \Theta_0$ vs. $\mathscr{H}_1 : \theta \in \Theta_1$.
  Here, $\mathscr{X} = \mathscr{X}_0 \cup \mathscr{X}_1$ with $\mathscr{X}_0 \cap \mathscr{X}_1 = \{\}$. As is well known, $\mathscr{X}_0 = \{x : \text{accept}\, \mathscr{H}_0 / \text{reject}\, \mathscr{H}_1\}$ is referred to as *acceptance region*, $\mathscr{X}_1 = \{x : \text{reject}\, \mathscr{H}_0 / \text{accept}\, \mathscr{H}_1\}$ as *critical (rejection) region*.
  Furthermore, an index of the the accepted hypothesis is given by the *test function*

$$\phi(x) = \begin{cases} 0 & x \in \mathscr{X}_0 \\ 1 & x \in \mathscr{X}_1 \end{cases}$$

- In case of $\Theta = \bigcup_{i=0}^{M-1} \Theta_i$ and $\bigcap_{i=0}^{M-1} \Theta_i = \{\}$ with $M > 2$, we speak about a *multiple (M-ary) hypothesis test*: $\mathscr{H}_0 : \theta \in \Theta_0$ vs. ... vs. $\mathscr{H}_{M-1} : \theta \in \Theta_{M-1}$.

**Definition 18** (Distinction via size of subsets). We distinguish two types of hypotheses:

- If $\Theta_i = \{\theta_i\}$ consists of a single element, $\mathscr{H}_i : \theta = \theta_i$ is called *simple hypothesis*.

- If $\Theta_i$ contains more than one element, $\mathscr{H}_i : \theta \in \Theta_i$ is called *composite hypothesis*.

In this study we will consider binary, composite hypothesis tests. Here, in case of a scalar parameter $\theta$, an order relation is possible, which allows to distinguish between two-sided tests (i.e. $\mathcal{H}_0 : \theta = \theta_0$ vs. $\mathcal{H}_1 : \theta \neq \theta_0$) and one-sided tests ($\mathcal{H}_0 : \theta \leq \theta_0$ vs. $\mathcal{H}_1 : \theta > \theta_0$). Note that in case of vector-valued parameters $\theta$ (as in matched subspace filtering in Subsection 3.5.3), we have to consider the vector norm for one-sided tests, i.e. $\mathcal{H}_0 : \|\theta\| \leq \theta_0$ vs. $\mathcal{H}_1 : \|\theta\| > \theta_0$.

We have the following decision scheme for binary tests:

**Definition 19** (Possible decisions for binary tests). We distinguish 4 cases:

- $\mathcal{H}_0$ is correctly accepted (*correct non-detection*), if $\theta \in \Theta_0, x \in \mathcal{X}_0$.
  The *acceptance probability* is given by $P_A = P\{\phi(x) = 0 \,|\, \mathcal{H}_o\}$.

- $\mathcal{H}_1$ is erroneously accepted (*type I error / false alarm / false positive*), if $\theta \in \Theta_0, x \in \mathcal{X}_1$.
  The *false alarm probability / size / significance level* is given by $P_{FA} = P\{\phi(x) = 1 \,|\, \mathcal{H}_o\}$, sometimes referred to as $\alpha$.

- $\mathcal{H}_1$ is correctly accepted (*correct detection*), if $\theta \in \Theta_1, x \in \mathcal{X}_1$.
  The *detection probability / power* is given by $P_D = P\{\phi(x) = 1 \,|\, \mathcal{H}_1\}$, sometimes referred to as $\beta$.

- $\mathcal{H}_0$ is erroneously accepted (*type II error / miss / false negative*), if $\theta \in \Theta_1, x \in \mathcal{X}_0$.
  The *miss probability* is given by $P_M = P\{\phi(x) = 0 \,|\, \mathcal{H}_1\}$

Obviously, $P_A + P_{FA} = 1$ and $P_D + P_M = 1$. Therefore, a hypothesis test is fully specified by the false alarm probability $P_{FA}$ and the power $P_D$. Plotting these two probabilities against each other yields the *Receiver Operator Characteristics (ROC)*, compare Fig. 3.5.1 for an illustration. An ideal detection would be characterized by the operating point $(P_{FA}, P_D) = (0, 1)$, which is not achievable in practical applications. Therefore, one has to typically choose a trade-off between *sensitivity* (high $P_D$) and *specificity*

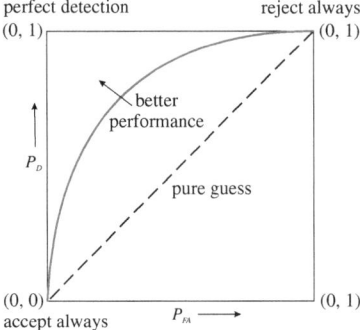

**Figure 3.5.1:** *Receiver Operator Characteristics. False alarm probability $P_{FA}$ and power $P_D$ plotted against each other for an imaginary test. Perfect detection would be achieved in the operating point $(P_{FA}, P_D) = (0, 1)$.*

(low $P_{FA}$) by adapting the test specifications (e.g. threshold) in such a way that the desired operating point in the ROC is attained.

### 3.5.2 Simple hypothesis testing

Although we do not address the topic of simple hypothesis testing in this study, we give a very brief overview in this subsection. This is intended as a basis for composite hypothesis testing discussed in Subsection 3.5.3.

As for simple binary tests $\Theta_i = \{\theta_i\}$, we have $p(x|\mathcal{H}_i) = p(x|\theta_i)$ for $i = \{0, 1\}$, i.e. the observed data bears information about the parameter and the hypothesis. This is different in composite hypothesis testing, see Subsection 3.5.3.

We would like to optimize both error probabilities, i.e. minimize $P_{FA}$ and maximize $P_D$. However, this is not possible at the same time, as changing $\phi(x)$ influences both probabilities. In simple hypothesis testing the *Neyman-Pearson criterion* is an alternative.

**Definition 20** (Neyman-Pearson Criterion)**.** Maximization of the power $P_D(\phi)$ while keeping the size below a prescribed level $\alpha$, i.e.

$$\phi_{NP}(x) \triangleq \arg\max_{\phi} P_D(\phi) \quad \text{subject to} \quad P_{FA}(\phi) \leq \alpha. \quad (3.5.1)$$

The Neyman-Pearson theorem then assures optimality, existence and uniqueness of an appropriate decision rule, compare Neyman and Pearson (1933). This case will be extended to composite hypothesis testing in the next subsection.

### 3.5.3 Composite hypothesis testing

In composite hypothesis tests there are more than one possible distributions of the observed data under a given hypothesis, as at least one partition of the parameter space consists of more than one element. Therefore, the mapping from a specific parameter $\theta \in \Theta_i$ to an observation $x$ is governed by a distribution depending on $\theta$, i.e. $p_\theta(x) = p(x; \theta)$. This makes composite hypothesis testing more difficult in comparison to simple hypothesis testing.

#### 3.5.3.1 Uniformly Most Powerful Tests

In our (non-Bayesian) setting, Neyman-Pearson theory has to be extended. As can be seen easily[13], the false alarm and detection probability of the decision rule $\phi(x)$ are given by

$$P_{FA}(\theta) = \mathbb{E}\{\phi(x); \theta\}, \theta \in \Theta_0$$
$$P_D(\theta) = \mathbb{E}\{\phi(x); \theta\}, \theta \in \Theta_1$$

and thus depend on $\theta$, as outlined above. Furthermore, we define the size $\alpha$ and the power $\beta(\theta)$ as

$$\alpha \triangleq \sup_{\theta \in \Theta_0} P_{FA}(\theta)$$
$$\beta(\theta) \triangleq P_D(\theta).$$

Now we consider a generalization of the Neyman-Pearson criterion (3.5.1) to composite hypothesis testing:

---

[13] We have (the analogous reasoning holds for $P_D(\theta)$)

$P_{FA}(\theta) = P\{\phi(x) = 1; \theta \in \Theta_0\} = \int_{x \in \mathscr{X}_1} \phi(x) p_\theta(x) dx + \underbrace{\int_{x \in \mathscr{X}_0} \phi(x) p_\theta(x) dx}_{=0} = \mathbb{E}\{\phi(x); \theta\}.$

**Definition 21** (Uniformly Most Powerful (UMP) Test)**.** A test $\phi(x)$ is uniformly most powerful of size $\alpha$, if its power is uniformly larger than the power of any other test of size $\alpha$, i.e.

$$\mathbb{E}\{\phi(x); \theta\} \geq \mathbb{E}\{\phi'(x); \theta\}, \theta \in \Theta_1$$

for all $\phi'(x)$ satisfying $\alpha = \sup_{\theta \in \Theta_0} \mathbb{E}\{\phi'(x); \theta\} \leq \alpha$.

Unfortunately it is often difficult to decide whether a UMP test exists for a specific composite hypothesis test, and if it does, to find the UMP decision rule.

However, in case of one-sided hypothesis tests with scalar parameters in the probability density function of the scalar observation, a positive existence statement can be made.

**Theorem 1** (Karlin-Rubin Theorem)**.** *If $x$ has monotone likelihood ratio, i.e. $L(x) = \frac{p(x;\theta_1)}{p(x;\theta_0)}$ is a non-decreasing function of $x$ for all pairs $\theta_1 > \theta_0$, then the level $\alpha$ UMP test for the problem $\mathcal{H}_0 : \theta \leq \theta_0$ vs. $\mathcal{H}_1 : \theta > \theta_0$ is given by*

$$\phi(x) = \begin{cases} 1 & x > \gamma \\ \eta & x = \gamma \\ 0 & x < \gamma \end{cases}$$

*where the randomization $\eta$ and the threshold $\gamma$ are chosen such that*

$$\mathbb{E}\{\phi(x); \theta_0\} = P\{x > \gamma; \theta_0\} + \eta\, P\{x = \gamma; \theta_0\} = \alpha.$$

*Proof.* Karlin and Rubin (1956). □

This theorem can be regarded as an extension of the Neyman-Pearson theorem to composite hypothesis testing and will be needed for the construction of matched subspace detectors, see below.

### 3.5.3.2 Invariance

A typical situation where it is impossible to find a UMP decision rule occurs in case of so-called *nuisance parameters*. For instance, let the parameter

$\theta$ consist of say $p$ components, but only $r < p$ components enter into the hypothesis. Therefore, it appears reasonable to restrict the test to a decision rule which is invariant of the $p - r + 1$ parameters not used and to find the most powerful invariant test.

Another motivation for invariance is the occurrence of symmetries. As an example, consider two-dimensional measurements $(x, y)$. If we want to test $\mathcal{H}_0 : D \leq D_0$ vs. $\mathcal{H}_1 : D > D_0$, with $D$ being the (average) distance from the origin, it seems natural that a decision rule should not depend on the exact location of $(x, y)$, but rather on the radius $\sqrt{x^2 + y^2}$. In this case a circular invariance would appear reasonable.

For a formalization consider the following definitions:

**Definition 22** (Invariance). Consider a family of probability density functions $\mathcal{P} = \{p(x; \theta), \theta \in \Theta\}$ and the binary composite hypothesis test $\mathcal{H}_0 : \theta \in \Theta_0$ vs. $\mathcal{H}_1 : \theta \in \Theta_1$. We describe symmetries/invariances by a (in algebraic sense) group $\mathcal{G}$ of transformations $g(x)$.

**Invariance of transformation groups:** We say $\mathcal{P}$ is invariant to $\mathcal{G}$ if, for any $x \in \mathcal{X}$ and $g \in \mathcal{G}$, $p_1(x_1; \theta)$ of $x_1 = g(x)$ can be written as

$$p_1(x_1; \theta) = p(x_1; \theta_1), \text{ for some } \theta_1 = \bar{g}(\theta).$$

The parameter transformation $\bar{g}$ is called the induced transformation on $\Theta$.

**Invariance of hypothesis testing problem:** If in addition $\bar{g}$ preserves the dichotomy $\Theta = \Theta_0 \cup \Theta_1$, i.e.

$$\theta \in \Theta_0 \Leftrightarrow \bar{g}(\theta) \in \Theta_0,$$
$$\theta \in \Theta_1 \Leftrightarrow \bar{g}(\theta) \in \Theta_1,$$

then we say that the hypothesis testing problem is invariant to $\mathcal{G}$.

**Invariance of decision rule:** If in addition we have, for any $x \in \mathcal{X}$ and $g \in \mathcal{G}$,

$$\phi(g(x)) = \phi(x),$$

we say that the test/decision rule $\phi(x)$ is invariant to $\mathscr{G}$.

Most powerful invariant tests are a restriction of uniformly most powerful tests to invariant decision rules:

**Definition 23** (Most powerful invariant test). The level $\alpha$ most powerful invariant test $\phi(x)$ is defined by the following conditions:
1. it has size $\alpha = \sup_{\theta \in \Theta_0} P_{FA}(\theta)$;
2. it is invariant to $\mathscr{G}$, i.e. $\phi(g(x)) = \phi(x)$;
3. its power is uniformly larger than the power of any other invariant test of size $\alpha$, i.e.

$$\mathbb{E}\{\phi(x); \theta\} \geq \mathbb{E}\{\phi'(x); \theta\}, \theta \in \Theta_1$$

for all $\phi'(x)$ satisfying $\phi'(g(x)) = \phi'(x)$, $\alpha = \sup_{\theta \in \Theta_0} \mathbb{E}\{\phi'(x); \theta\} \leq \alpha$.

In order to find most powerful invariant tests, we will need the concept of maximal invariant statistics.

**Definition 24** (Maximal invariant statistic). A statistic $T(x)$ is said to be maximal invariant if it satisfies
1. $T(g(x)) = T(x)$ for all $g \in \mathscr{G}$;
2. $T(x_1) = T(x_2)$ implies $x_2 = g(x_1)$ for some $g \in \mathscr{G}$.

Thus, a maximal invariant statistic is constant on the orbits of $\mathscr{G}$, i.e. on $\{g(x) : g \in \mathscr{G}\}$, but takes a different value for each orbit. The following theorem establishes a link between maximal invariant statistics and invariant decision rules.

**Theorem 2** (Invariance of decision rule). Let $T(x)$ be a maximal invariant statistic with respect to the transformation group $\mathscr{G}$.

Then $\phi(x)$ is an invariant decision rule iff it depends on x only through $T(x)$, i.e. $\phi(x) = \phi_I(x)$ for all x and some function $\phi_I(x)$, or equivalently

$$T(x_1) = T(x_2) \Rightarrow \phi(x_1) = \phi(x_2).$$

*Proof.* Appendix. □

Thus, summing up, the following approach has to be taken in order to solve a composite hypothesis problem using invariance principles:
1. Identify a meaningful transformation group $\mathscr{G}$.
2. Check the invariance of the problem with respect to $\mathscr{G}$.
3. Find a maximal invariant statistic $T(x)$.
4. Devise a UMP detector for the problem based on $T(x)$, e.g. by application of the Karlin-Rubin theorem 1.

### 3.5.3.3 Matched subspace filters

Matched subspace detectors are an application of the invariance principle to composite hypothesis testing, compare the review article by Scharf and Friedlander (1994).

Consider the linear data model

$$\mathbf{x} = \mathbf{s} + \mathbf{v} + \mathbf{w}, \quad \mathbf{s} \in \mathscr{S}, \mathbf{v} \in \mathscr{S}^\perp, \mathbf{w} \sim \mathcal{N}(\mathbf{0}, \sigma^2 \mathbf{I}) \quad (3.5.2)$$

where the involved entities are $N \times 1$ vectors (i.e. signal samples have been stacked into a vector). We assume that the signal consists of a linear combination of modes, i.e. $\mathbf{s} = \mathbf{H}\boldsymbol{\theta}$ where $\mathbf{H}$ is a known modal matrix of dimension $N \times p$ and $\boldsymbol{\theta}$ an unknown parameter vector (mode weights) of dimension $p \times 1$. Thus, $\mathbf{s}$ lies in the subspace $\mathscr{S}$ spanned by the columns of $\mathbf{H}$, compare Fig. 3.5.2. However, we do not know the direction of $\mathbf{s}$ within $\mathscr{S}$. Furthermore, $\mathbf{v} \in \mathscr{S}^\perp$ is an unknown influence in the orthogonal subspace to $\mathscr{S}$, and $\mathbf{w}$ is additive noise. Thus, $\mathbf{x} \sim \mathcal{N}(\mathbf{H}\boldsymbol{\theta} + \mathbf{v}, \sigma^2 \mathbf{I})$.

The goal is to test whether the signal $\mathbf{s} \in \mathscr{S}$ is present in the measured data $\mathbf{x}$ or not, i.e. $\mathcal{H}_0 : \|\boldsymbol{\theta}\| = 0$ vs. $\mathcal{H}_1 : \|\boldsymbol{\theta}\| > 0$, as illustrated in Fig. 3.5.2. This is a one-sided composite hypothesis test for which no UMP decision rule exists ($\boldsymbol{\theta}$ and $\mathbf{v}$ are both unknown, but the latter is a nuisance parameter). Therefore, we reduce the problem by application of invariance principles:

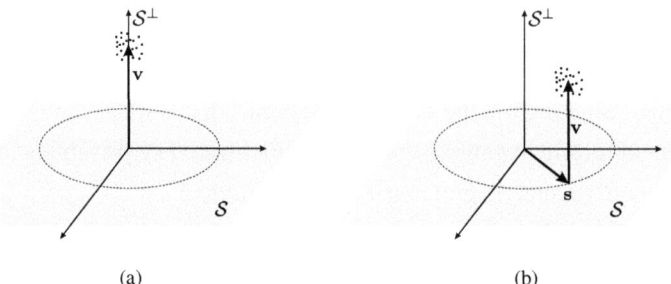

**Figure 3.5.2: Symmetry of linear data model.** In case of (a) $\mathcal{H}_0 : \|\theta\| = 0$ the vector **s** vanishes, in case of (b) $\mathcal{H}_1 : \|\theta\| > 0$ the vector $\mathbf{s} = \mathbf{H}\theta$ lies in $\mathcal{S}$. In both cases the vector **v** represents an offset in the direction of the orthogonal subspace $\mathcal{S}^\perp$.

1. First, we identify a meaningful transformation group. As the direction of $\mathbf{s} \in \mathcal{S}$ and $\mathbf{v} \in \mathcal{S}^\perp$ are unknown, it seems natural to demand invariance with respect to rotation within $\mathcal{S}$ and a bias in $\mathcal{S}^\perp$. This can be represented by the transformation

$$g : \mathbf{x}_1 = g(\mathbf{x}) = \mathbf{Q}_\mathcal{S}\mathbf{x} + \mathbf{z}, \quad \mathbf{z} \in \mathcal{S}^\perp,$$

where the rotation matrix is $\mathbf{Q}_\mathcal{S} = \mathbf{U}\mathbf{Q}\mathbf{U}^T + \mathbf{P}_{\mathcal{S}^\perp}$. Hereby, **U** is a $N \times p$ matrix whose columns constitute an orthogonal basis of $\mathcal{S}$, **Q** is an arbitrary $p \times p$ orthogonal matrix and $\mathbf{P}_{\mathcal{S}^\perp} = \mathbf{I} - \mathbf{U}\mathbf{U}^T$ is the orthogonal projection on $\mathcal{S}^\perp$. Compare Fig. 3.5.3 (a) for an illustration.

Thus, $\mathcal{G} = \{g : g(\mathbf{x}) = \mathbf{Q}_\mathcal{S}\mathbf{x} + \mathbf{z}\}$ as detailed above. This set of affine transformations obviously satisfies algebraic group properties. In particular $\mathbf{Q}_\mathcal{S}\mathbf{x} + \mathbf{z} = \mathbf{Q}_\mathcal{S}(\mathbf{x} + \mathbf{z})$, as $\mathbf{z} \in \mathcal{S}^\perp$, compare Fig. 3.5.3 (a).

2. It can be easily seen that the hypothesis testing problem is invariant with respect to $\mathcal{G}$, compare the annex.

3. The next step is to find a maximal invariant test statistic $T(\mathbf{x})$. Note that here the orbits of $\mathcal{G}$ are the hyper-cylinders $\mathcal{C}_r = \{\mathbf{x} : \|\mathbf{P}_\mathcal{S}\mathbf{x}\|^2 =$

$r$, $\mathbf{P}_{\mathscr{S}^\perp}\mathbf{x}$ arbitrary}, with $\mathbf{P}_{\mathscr{S}} = \mathbf{U}\mathbf{U}^T$ denoting the orthogonal projection on $\mathscr{S}$ (compare Fig. 3.5.3 (a) for an illustration). Therefore, a candidate is $T(x) = \|\mathbf{P}_{\mathscr{S}}\mathbf{x}\|^2$, and short calculations (compare the annex) show that this statistic is in fact maximal invariant.

Note that $T(\mathbf{x})$ measures the energy of the signal in the subspace $\mathscr{S}$.

4. In order to find the most powerful invariant test, we therefore restrict to decision rules $\phi(x)$ depending only on $T(x) = \|\mathbf{P}_{\mathscr{S}}\mathbf{x}\|^2$, according to theorem 2. As $T(\mathbf{x}) = \mathbf{x}^T \mathbf{P}_{\mathscr{S}} \mathbf{x}$ is the sum of $p$ squares, the quotient

$$\chi^2(\mathbf{x}) = \frac{\mathbf{x}^T \mathbf{P}_{\mathscr{S}} \mathbf{x}}{\sigma^2} \tag{3.5.3}$$

is $\chi^2$-distributed (non-centrally with $p$ degrees of freedom).[14] Consequently, as the $\chi^2$-distribution has non-decreasing likelihood ratio, the Karlin-Rubin theorem 1 implies that the most powerful invariant test is

$$\phi(\mathbf{x}) = \phi_I(\chi^2(\mathbf{x})) = \begin{cases} 1 & \chi^2(\mathbf{x}) > \gamma \\ \eta & \chi^2(\mathbf{x}) = \gamma, \\ 0 & \chi^2(\mathbf{x}) < \gamma \end{cases}$$

with $\eta$ and $\gamma$ chosen such that $\mathbb{E}\{\phi(\mathbf{x}); \|\theta\| = 0\} = \alpha$.

This decision rule is called *matched subspace detector*, because $\mathbf{x}$ is filtered by $\mathbf{P}_{\mathscr{S}}$, i.e. a filter is matched to the subspace $\mathscr{S}$ under consideration.

However, in practical applications the noise variance $\sigma^2$ is most likely unknown in model 3.5.2. Therefore, there is no way to compute the statistic $\frac{\mathbf{x}^T \mathbf{P}_{\mathscr{S}} \mathbf{x}}{\sigma^2}$, consequently there is no UMP test which is invariant to rotation and orthogonal bias.

The principal approach for deriving a UMP invariant detector is analogous to the case with known variance, though more technical (Scharf and Lytle 1971). Therefore, we only give a quick overview, also compare Scharf (1991).

---

[14] Note that $\chi^2(\mathbf{x})$ differs from $T(\mathbf{x})$ only by the scalar constant factor $\sigma^2$, thus theorem 2 still applies.

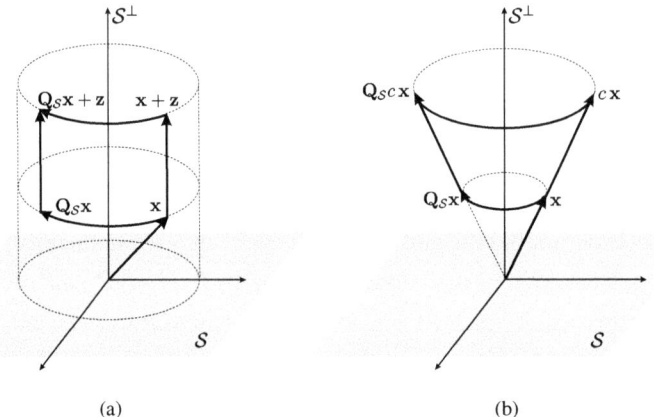

**Figure 3.5.3:** *Matched subspace filter. Illustration of the invariance properties. (a) In case of known noise variance $\sigma^2$ the orbits of the transform group $\mathscr{G}$ represent a hyper-cylinder, (b) in case of unknown noise variance $\sigma^2$ a hyper-cone.*

1. Here we ask for invariance with respect to transformations which scale the measurements and rotate them in the signal subspace $\mathscr{S}$. This can be represented by the transformation

$$g : \mathbf{x}_1 = g(\mathbf{x}) = \mathbf{Q}_{\mathscr{S}} c \mathbf{x},$$

   where $c$ is a scaling factor, and the rotation matrix is again $\mathbf{Q}_{\mathscr{S}} = \mathbf{U}\mathbf{Q}\mathbf{U}^T + \mathbf{P}_{\mathscr{S}^\perp}$. Compare Fig. 3.5.3 (b) for an illustration.
   Thus, $\mathscr{G} = \{g : g(\mathbf{x}) = \mathbf{Q}_{\mathscr{S}} c\mathbf{x}\}$, which can be shown to satisfy algebraic group properties.

2. Again, the problem is invariant to $\mathscr{G}$.

3. A maximal invariant statistic is given by $T(\mathbf{x}) = \frac{\mathbf{x}^T \mathbf{P}_H \mathbf{x}}{\mathbf{x}^T (\mathbf{I} - \mathbf{P}_H) \mathbf{x}}$. Here, the orbits of $T(\mathbf{x})$ form hyper-cones, compare Fig. 3.5.3 (b).

4. As before, we restrict to decision rules $\phi(\mathbf{x})$ only depending on $T(\mathbf{x})$. A slight modification of the maximal invariant statistic leads to

$$F = \frac{\mathbf{x}^T \mathbf{P}_H \mathbf{x} / \sigma^2 p}{\mathbf{x}^T (\mathbf{I} - \mathbf{P}_H) \mathbf{x} / \sigma^2 (N-p)}, \qquad (3.5.4)$$

which again differs from $T(\mathbf{x})$ only by a constant scalar factor. Thus, theorem 2 applies.

Statistic (3.5.4) is a ratio of quadratic forms in projection matrices. Therefore, it measures the ratio of the energy of $\mathbf{x}$ in the subspace $\mathscr{S}$ to the energy of $\mathbf{x}$ in the subspace $\mathscr{S}^\perp$ (per dimension). Compare Fig. 3.5.3 (b) for an illustration.

As $F$ is a ratio of independent $\chi^2$-distributed random variables (quadratic forms), it is $F$-distributed (non-centrally with $p$ and $N-p$ degrees of freedom). Again, as the $F$-distribution has monotone likelihood ratio, the Karlin-Rubin theorem 1 applies. Thus, the most powerful invariant test is given by

$$\phi(\mathbf{x}) = \phi_I(F(\mathbf{x})) = \begin{cases} 1 & F(\mathbf{x}) > \gamma \\ \eta & F(\mathbf{x}) = \gamma, \\ 0 & F(\mathbf{x}) < \gamma \end{cases}$$

with $\eta$ and $\gamma$ chosen such that $\mathbb{E}\{\phi(\mathbf{x}); \mu = 0\} = \alpha$.

This decision rule is called *CFAR matched subspace detector*, because it has constant false alarm rate independently of $\sigma^2$. [15]

## 3.6 Factor Models

Factor models in a time series setting may be used to compress information contained in the data in both the cross-sectional dimension, $N$ say, and in the time dimension $T$. In this way it is possible to overcome the »curse of dimensionality« plaguing traditional multivariate time series modeling [...].

— Deistler et al. (2010): *Generalized Dynamic Factor Models: An Approach via Singular Autoregressions*, recent theoretical considerations on factor models

The use of factor analysis was established by psychologists at the beginning of the 20th century for explaining common determinants of intel-

---
[15]The division by $\sigma^2$ cancels out in the numerator and denominator of (3.5.4); otherwise the statistic could not be computed anyway.

ligence (Burt 1909). Factor models in the time series context exploit this idea of common factors in case of a chronological order of the observations (Geweke 1977). They are often used for the analysis and forecasting of high-dimensional signals, when their single components show similarities or a kind of co-movement. This modeling approach is in particular popular in EEG recordings (Molenaar 1985, Molenaar and Nesselroade 2001).

We refer to Anderson (2003) for factor analysis and to Bartholomew et al. (2011) and Loehlin (2004) for an introduction to latent variable models, in particular factor models.

Let us consider a $K$-dimensional signal $\mathbf{x}[n]$ with components $x_k[n]$, $k = 1,...,K$ which is weakly stationary with mean zero. The basic idea of factor models is to separate this observed signal into a part representing the co-movement and a part representing the individual movements (»noise«) of the data. This idea is summarized in the following:

**Definition 25** (Factor Model). In its general form, a factor model is written as
$$\mathbf{x}[n] = \chi[n] + \eta[n], \qquad (3.6.1)$$
with the *latent variables* $\chi[n]$ and the noise $\eta[n]$.

The $K$-dimensional latent variables $\chi[n]$ are generated by a $q$-dimensional process, where $q \ll K$. These $q$ driving processes are called *factors*, therefrom the term *factor model*.

This separation into latent variables and noise can be achieved by different means. In this study we will separate the latent variables $\chi[n]$ from the noise by means of *Principal Component Analysis (PCA)* and derive the *static factors* $\mathbf{z}[n]$. See Pearson (1901), Hotelling (1933) and Jackson (2004) for background information on PCA.

In this study we consider a specific class of factor models which we will need in Chapter 7. Following PCA we model the static factors $\mathbf{z}[n]$ as a regular AR(p) process. This modeling approach is sometimes referred to as a *quasi-static factor model*, see Deistler and Zinner (2007) for theoretical considerations. Here we limit ourselves to a summary of the implementation:

**Definition 26.** Construction of the quasi-static factor model.

1. In order to separate $\mathbf{x}[n] = \chi[n] + \eta[n]$, we apply the PCA: We calculate the eigenvalue decomposition of the covariance matrix of the observed signal

$$\begin{aligned}\operatorname{Cov}\{\mathbf{x}[n]\} &= \mathbf{O}\Gamma\mathbf{O}^T \\ &= (\mathbf{O}_1\ \mathbf{O}_2) \begin{pmatrix} \Gamma_1 & 0 \\ 0 & \Gamma_2 \end{pmatrix} \begin{pmatrix} \mathbf{O}_1^T \\ \mathbf{O}_2^T \end{pmatrix} \\ &= \mathbf{O}_1\Gamma_1\mathbf{O}_1^T + \mathbf{O}_2\Gamma_2\mathbf{O}_2^T,\end{aligned}$$

where $\Gamma = \operatorname{diag}(\lambda_1,\ldots,\lambda_n)$ contains the ordered eigenvalues $\lambda_1 \geq \lambda_2 \geq \ldots \geq \lambda_n > 0$, and $\mathbf{O}$ is an orthogonal matrix. $\mathbf{O}_1$ contains the first $q$ columns of $\mathbf{O}$ corresponding to the $q$ largest eigenvalues, respectively $\mathbf{O}_2$ the remaining $n-q$ columns.
Let

$$\hat{\mathbf{z}}[n] = \mathbf{O}_1^T \mathbf{x}[n]$$

be the estimator of the $q$-dimensional ($q \ll K$) *static factors*. Then the estimator of the $K$-dimensional latent variables is obtained by

$$\hat{\chi}[n] = \mathbf{O}_1\hat{\mathbf{z}}[n] = \Lambda\hat{\mathbf{z}}[n], \tag{3.6.2}$$

where $\Lambda$ is termed *factor loading matrix*.

As is well known, one could generate other static factors by premultiplication with a regular (in particular orthogonal) matrix $\mathbf{U}^{16}$, i.e.

$$\chi[n] = \Lambda\mathbf{z}[n] = \underbrace{\Lambda\mathbf{U}^{-1}}_{\tilde{\Lambda}}\underbrace{\mathbf{U}\mathbf{z}[n]}_{\tilde{\mathbf{z}}[n]} = \tilde{\Lambda}\tilde{\mathbf{z}}[n].$$

---

[16] However, this will not impair our causality analysis in Chapter 7, see Flamm et al. (2013).

2. We model the static factors $\mathbf{z}[n]$ as a regular AR(p) process, compare Section 3.2. According to the compact notation (3.2.2) we write

$$\mathbf{A}(z)\mathbf{z}[n] = \varepsilon[n], \qquad (3.6.3)$$

where $\mathbf{A}[0] = \mathbf{I}_{q \times q}$ and $\Sigma_\varepsilon > 0$ as in model (3.2.1). Furthermore we assume the stability condition (3.2.3) to hold.

3. Equations (3.6.2), (3.6.3) and the causal invertibility of $\mathbf{A}(z)$ according to (3.2.3) together yield

$$\chi[n] = \Lambda \mathbf{z}[n] = \Lambda \mathbf{A}^{-1}(z)\varepsilon[n]. \qquad (3.6.4)$$

In other words the $q$-dimensional white noise $\varepsilon[n]$ generates the $K$-dimensional latent variables $\chi[n]$.

# Part II

# Materials and Methods

# 4 Propagation Analysis Framework

> La filosofia è scritta in questo grandissimo libro che continuamente ci sta aperto innanzi a gli occhi (io dico l'universo), ma non si può intendere se prima non s'impara a intender la lingua, e conoscer i caratteri, ne' quali è scritto. Egli è scritto in lingua matematica, e i caratteri son triangoli, cerchi, ed altre figure geometriche, senza i quali mezi è impossibile a intenderne umanamente parola; senza questi è un aggirarsi vanamente per un oscuro laberinto.
>
> — Galileo Galilei: *Il Saggiatore*
>
> Philosophy is written in this grand book — I mean the universe — which stands continually open to our gaze, but it cannot be understood unless one first learns to comprehend the language and interpret the characters in which it is written. It is written in the language of mathematics, and its characters are triangles, circles, and other geometrical figures, without which it is humanly impossible to understand a single word of it; without these, one is wandering around in a dark labyrinth.
>
> — *The Assayer*[1] by Galileo Galilei (1564-1642), who played a major role in the scientific revolution in Renaissance era

## 4.1 Introduction

In this chapter we discuss the central point of this study, a framework for epileptic seizure propagation analysis.

This section serves as an introduction and is dedicated to the medical and technical background of propagation analysis. We elaborate our framework in Section 4.2 and present patient history and recorded data in Section 4.3.

---

[1]Chapter 6; English translation by Stillman Drake.

### 4.1.1 Medical background

The analysis of seizure propagation in the course of a presurgical evaluation is a clinically relevant task: Determination of the SOZ is a prerequisite for a surgical intervention, and the characteristics of the early seizure spread yield valuable information about the expected post-surgical outcome: As mentioned in Subsection 2.3.6, a poor postsurgical outcome is highly correlated with contralateral propagation, see e.g. the surface-EEG study by Schulz et al. (2000). Furthermore, it is associated with fast propagation, see e.g. the studies based on depth electrodes by Lieb et al. (1986) or on ECoG by Weinand et al. (1992).

In this study we are concerned with the propagation of epileptic activity in ECoG. While propagation can be easily defined in surface EEG in terms of ipsi- vs. contralateral due standardized recording schemes (see Subsection 2.2.2), this analysis is more difficult in intracranial EEG due to the patient-specific electrode topology. In return the increased spatial resolution allows for more detailed insights into seizure spread.

Recently two clinical studies based on intracerebral recordings shed light onto early seizure spread.

First, Götz-Trabert et al. (2008) analyzed the electrographic onset, the initial propagation of epileptic activity and the delay to the clinical onset. Hereby, propagated activity was defined as clearly identifiable ictal patterns at electrode positions of at least 2 cm in distance from its origin, i.e. the SOZ. They found that in case of mTLE activity initially propagated in basal temporal direction, with an average delay of 13.7 seconds from electrographic onset to initial propagation and 26.8 seconds to clinical onset. In neocortical TLE (nTLE) propagation was faster (average delay to initial propagation of 7.6 seconds, to clinical onset of 17.7 seconds), and in case of frontal original even faster (average delay to initial propagation of 3.8 seconds, to clinical onset of 10.6 seconds).

Second, Jenssen et al. (2011) reported that seizure propagation varied depending on the localization of the SOZ. For instance, one frequent pathway led from mesial temporal to contralateral mesial temporal structures. Moreover, the observed ictal onset frequency was higher than the frequency of propagated activity.

### 4.1.2 Technical background

While seizure spread is analyzed by visual inspection of the raw EEG recordings in a clinical environment, we pursue another approach: propagation analysis based on signal processing techniques. This computer-based methodology allows to avoid the aforementioned time-consuming manual analysis of ictal patterns, which is highly subjective and depends on the individual experience of the examiner.

Considerations to interpret the EEG as a signal with statistical properties and to apply signal processing techniques date back for more than 80 years (Dietsch 1932). The interest in computational methods of EEG recordings significantly increased with the early rise of modern computers, see e.g. Gasser (1979). Throughout the last decades neuroscience has evolved as an interdisciplinary science branch uniting methodology from informatics, physics, statistics and medicine. This diversity is reflected by such different approaches as machine learning (Formisano et al. 2008), structural equation modeling (Astolfi, Cincotti, Mattia, Lai, Baccala, de Vico Fallani, Salinari, Ursino, Zavaglia and Babiloni 2005), hidden Markov models (Cassidy and Brown 2002), dynamic causal modeling (Friston et al. 2003) or time-variant signal processing (Hesse et al. 2003).

In this study we consider seizure propagation, which comprises the localization of the SOZ and the determination of early seizure spread, given the exact starting time of the epileptic activity. Note the distinction from three similar topics in quantitative EEG analysis: The goal of 3D source localization is the determination of the source of epileptic activity within the brain (i.e. somewhere between the electrode positions in 3D space), see e.g. the review papers by Plummer et al. (2008) and Grech et al. (2008). In seizure detection one aims at the automatic detection of seizures without any a-priori time information, see e.g. the introduction by Gotman (2003). Finally, the idea of seizure prediction is to predict a seizure from changes in the preictal state of the EEG, compare e.g. the review by Mormann et al. (2007).

For our purposes we stick to classical linear methodology involving non-parametric spectral estimation, autoregressive modeling and matched subspace filtering (compare Chapter 3).

Recent approaches to epileptic focus detection and initial seizure propagation analysis of invasive EEG data include the use of time-variant dependency measures via Kalman filtering (van Mierlo et al. 2011), 3D source localization with subsequent coupling quantification (Kim et al. 2010), dependency analysis in the context of classical AR modeling (Ge et al. 2007) or cross-correlation studies (Mizuno-Matsumoto et al. 1999).

We do not make use of non-linear methodology, although recently Hegde et al. (2005) and Andrzejak et al. (2006) and very recently Papana and Kugiumtzis (2012) have successfully applied non-linear approaches to epileptic invasive EEG. In the last years, however, the trend seems to switch back to linear frameworks, compare Blinowska (2011): This shift is due to the simplicity of the well-established linear methods, in particular autoregressive modeling, and the error-proneness of the non-linear methods requiring phase space embedding (dimension to be chosen without any direct physical interpretation, long stationary data segments needed). In particular this physical interpretability makes linear methodology so popular.

In order to identify the SOZ and initial seizure spread we look at the first few, typically four, seconds of epileptic activity (see the next subsection). During this period we try to capture the dynamics imposed by the focus, whereas in later periods any causal information is lost due to the common rhythmic behavior resulting out of feedback-processes (Zoldi et al. 2000).

While it is commonly agreed on that there are differences between interictal and ictal periods (demonstrated e.g. via non-linear methodology by Geng and Zhou (2010)[2] or nuclear medical imaging by Baumgartner et al. (1998)), the exact mechanisms have not been conclusively revealed yet: Jouny et al. (2003, 2010) claimed that signal complexity rises before focal seizures, whereas Mormann et al. (2003) reported the opposite result. Furthermore, according to Lehnertz and Elger (1995), the signal complexity decreases in the course of a seizure, in particular in vicinity of the focus. This hypothesis of decreasing neuronal synchronization is supported by a study in experimental setup (Netoff and Schiff 2002), showing evidence that synchronicity may disrupt epileptic activity. In accordance with this argument (but in contrast to Lehnertz and Elger (1995)), Schindler, Leung, Elger and Lehnertz (2007) and Schindler, Elger and Lehnertz (2007) claimed that synchronicity increases at the end of a seizure, which they think of as a cerebral mechanism for seizure termination. Finally, Warren et al. (2010) reported lower synchronicity between the seizure generating region and other brain areas, thus a some-how isolated SOZ.

In our point of view, these discrepancies reflect the wide variety of seizure generating mechanisms in different patients.

Finally note that stationarity issues are not in the scope of this study. Assuming short-term stationarity, we employ short data windows throughout this study and apply classical methodology for stationary processes within these short data windows. We refer to the diploma thesis of Graef (2008) for stationarity analysis and time-variant autoregressive modeling of ECoG. Tests for stationarity analysis in neuroscience are reviewed in Kipinski et al. (2011).

## 4.2 Framework

In this section we discuss the central point of this study, a framework for epileptic seizure propagation analysis. In our context, seizure propagation analysis comprises two aspects: the determination of the early seizure

---

[2]Compare Fusheng et al. (2000) for an explanatory attempt of the physical interpretability of *Approximate Entropy*.

spread (»Where does the activity propagate to?«), and as a sub-problem the determination of the seizure onset zone (»Where does this activity originate from?«). This analysis is based on the localizing value of ictal (and interictal) patterns in invasive EEG, compare Subsection 2.3.5

In order to analyze seizure propagation, we compare four different approaches, see Table 4.2.1.

First, the detection of ictal HFOs allows to determine the HFO-generating zone and the initial propagation of HFOs. Initial ictal HFOs typically precede conventional ictal patterns by several seconds (Khosravani et al. 2009, Imamura et al. 2011), and the HFO-generating zone is highly correlated with the SOZ (Usui et al. 2011). This approach is detailed in Chapter 5.

Second, the application of causality measures in the context of autoregressive modeling allows to determine the SOZ. The initial spread of hyper-synchronous epileptic activity is indicated by arrows, which point away from the SOZ. For technical reasons, a reduction of the number of channels is needed for appropriate estimation of the AR model. We propose two different automatic methods: a channel selection algorithm in Chapter 6 and factor models in Chapter 7.

Third, segmentation of the individual channels and classification of the segments regarding their epileptic character yields the SOZ and the seizure spread. The temporal delay of the start of epileptic activity on different channels is an indicator for seizure propagation. The channels showing epileptic activity first mark the SOZ. This approach is pursued in Chapter 8.

Forth, we compare the results of the above methods with the clinical findings which represent the current gold standard in presurgical evaluation. Here, visual analysis of interictal and ictal EEG patterns, symptomatology, diagnostic imaging techniques and neuropsychological tests are considered. This comparison is performed in Chapter 9 and discussed in Chapter 10.

## Seizure propagation analysis with sub-problem SOZ detection

|  | HFOs | Causality measures | Segmentation | Clinical findings |
|---|---|---|---|---|
| **Idea** | – Detection of ictal HFOs on individual channels and determination of their respective propagation | – Representation of hyper-synchronous activity via arrows<br>– Dimension reduction via channel selection or factor model (for multivariate AR modeling)<br>– Arrows point away from the electrodes of the SOZ | – Segmentation of individual channels and classification of the epileptogenic character of the segments<br>– Delay of epileptogenic segments on different channels indicates SOZ and propagation | – Visual analysis of ECoG (ictal patterns, interictal spikes)<br>– MRI (lesion?)<br>– Symptomatology<br>– Neuropsychological testing<br>– If possible: classification of postsurgical outcome (Engel) |
| **Clin. relev.** | – Good correlation of HFOs with SOZ | – Direct indication of SOZ (by arrows) | – Direct indication of SOZ (leading epileptogenic segments)<br>– Determination of early seizure spread | – Directly relevant |

**Table 4.2.1:** *Seizure propagation analysis framework detailing approaches and their respective clinical relevance. The framework is built on four technical methods and the clinical findings.*

## 4.3 Data

### 4.3.1 Patient history

The ECoG data used in this study are taken from a patient (male, 44 years) who has been suffering from therapy-resistant focal epilepsy since the age of twelve. Apart from a short seizure-free period in adolescence there were no seizure-free intervals in the last years. According to history, seizures last from 20 to 60 seconds, and seizure frequency shows a wide range between few seizures per week and several ones per day. Seizures manifest as motionless-stare with impairment of consciousness and may be followed by tonic-clonic seizures. During the last months before the reported examination he typically encountered one to five seizures per day.

A first prolonged video-EEG monitoring in spring 2009 (using surface electrodes) confirmed the diagnosis of MRI-negative right-hemispheric focal epilepsy, but a prescribed localization of the epileptic focus was not possible in this setting, compare Section 10.2.1.

### 4.3.2 ECoG recordings

Consequently, the patient was admitted for prolonged invasive video-EEG monitoring in fall 2011. Three subdural strip electrodes with a total of 25 channels were implanted at Vienna General Hospital, University Clinic for Neurosurgery (see Fig. 4.3.1 for the electrode positions on the cortex). Subsequently the patient was transferred to the Epilepsy Monitoring Unit at Neurological Center Rosenhügel where he underwent a four-day-lasting video-EEG recording. Recording was performed at a sampling frequency of 1024 Hz using Micromed SystemPlus Evolution®, and electrode B1 was chosen as reference.

After reduction of anti-epileptic combination therapy (clobazam from 20 to 5 mg/day, lacosamide from 400 to 0 mg/day, oxcarbazepine from 1800 to 300 mg/day) on the 4th day, four seizures were registered. Due to decreasing data quality we only analyze the first three, see Figs. 4.3.2, 4.3.3

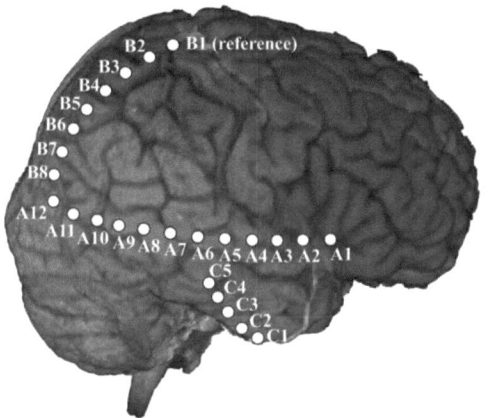

**Figure 4.3.1:** *MRI scan with electrode positions. Recordings are referenced to electrode B1 outside the seizure focus.*

and 4.3.4. Each seizure starts with paroxysmal fast activity of 30 Hz, in case of seizures 1 and 2 with antecedent HFOs (Figs. 4.3.2 and 4.3.3). After an intermediate phase (of average duration of 10 seconds[3]), we observe propagation of hyper-synchronous rhythmic activity in the $\vartheta/\alpha$-band.

After recording, the ECoG data were preprocessed in Matlab®: Line interference was removed using a notch filter at 50 Hz, and a high-pass filter at 1 Hz was applied in order to get rid of low-frequency contributions we are not intested in[4]. Except of the HFO analysis (see Chapter 5), the signals were low-pass filtered at 64 Hz in order to avoid aliasing and then downsampled to 128 Hz sampling rate.

Note that only in case of HFO detection we examine the initial electrographic onset, see Chapter 5. In all other approaches the onset of hyper-synchronous rhythmic $\vartheta$-activity is the relevant one, compare Chapter 6 for causality analysis, Chapter 7 for the influence analysis and Chapter 8 for the segmentation method.

---

[3]This is in good accordance with literature (Walczak et al. 1992).
[4]Compare Subsection 2.3.5 for ictal slow shifts.

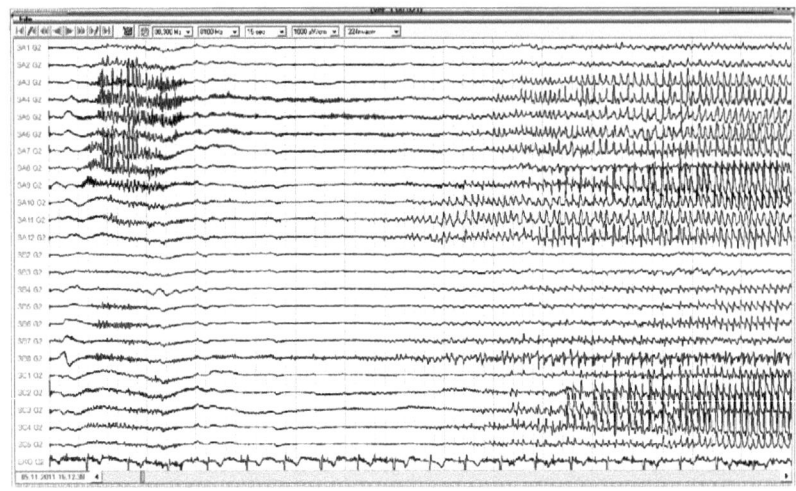

**Figure 4.3.2:** *Seizure 1, initial 15 seconds displayed. Electrographic onset at 16:12:38 with HFOs followed by paroxysmal fast activity of 30 Hz. Propagation of rhythmic $\vartheta$-activity starts at 16:12:45.*

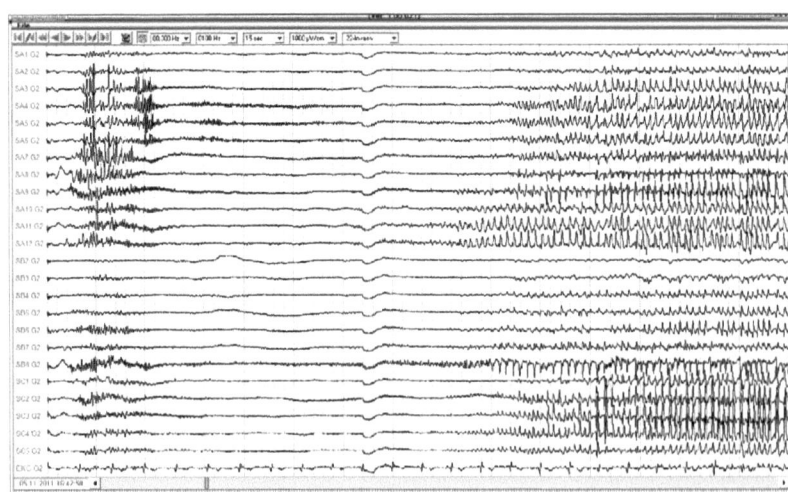

**Figure 4.3.3:** *Seizure 2, initial 15 seconds displayed. Electrographic onset at 16:47:88 with HFOs followed by paroxysmal fast activity of 30 Hz. Propagation of rhythmic $\vartheta$-activity starts at 16:48:06.*

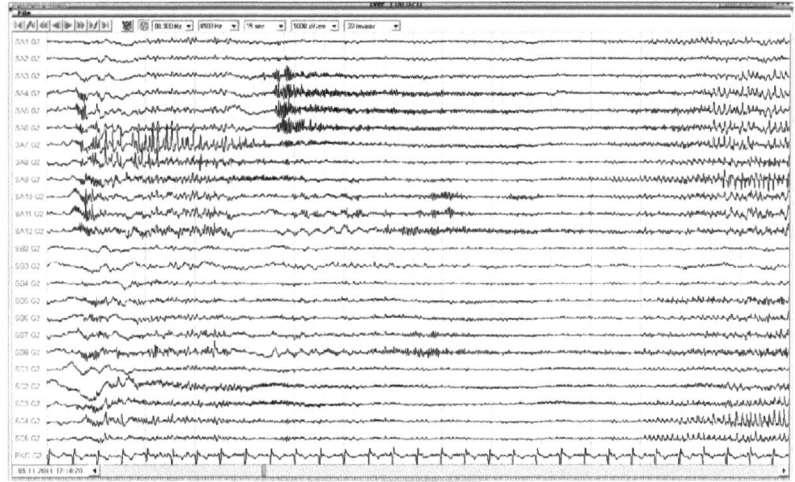

**Figure 4.3.4:** *Seizure 3, initial 15 seconds displayed.* Electrographic onset at 17:18:20 with paroxysmal fast activity of 30 Hz. Propagation of rhythmic $\vartheta$-activity starts at 17:18:32.

### 4.3.3 Visual inspection

Three clinical experts independently performed a visual inspection of the three seizures. They analyzed the propagation of high frequency oscillations (HFOs) as well as the one of conventional $\vartheta$-activity. In both cases they classified the involved electrodes into initial ones and close follow-up electrodes. The results of this analysis are summarized in Table 4.3.1 and will be used for assessment of the proposed technological methods.

| Seizure | Investigator | Initial Electrodes | Close follow-up |
|---|---|---|---|
| 1 | Expert 1 | B8, B7 | A11, A5, A6, A7, A12 |
|   | Expert 2 | B8, B7 | A11, A5, A6, A7, A12 |
|   | Expert 3 | B8, B7 | A5, A7, A11 |
| 2 | Expert 1 | A9 | C3 |
|   | Expert 2 | A9 | C3, C2 |
|   | Expert 3 | A9 | C3 |
| 3 | Expert 1 | - | - |
|   | Expert 2 | - | - |
|   | Expert 3 | - | - |

(a)

| Seizure | Investigator | Initial Electrodes | Close follow-up |
|---|---|---|---|
| 1 | Expert 1 | B8 | A10, A11, A12 |
|   | Expert 2 | A11, A12, B8 | A9, A10, B7 |
|   | Expert 4 | A10, A11, A12 | B8 |
| 2 | Expert 1 | A11, A12 | A9, A10 |
|   | Expert 2 | A11, A12 | A10 |
|   | Expert 4 | A11, A12 | B8 |
| 3 | Expert 1 | A9, A10 | A8, A11, A12, B6, B7, B8, C1, C4, C5 |
|   | Expert 2 | A9 | A1, A2, A3, C2, C3 |
|   | Expert 4 | A8, A9 | A1, C3, C4, C5 |

(b)

**Table 4.3.1:** *Visual inspection of the ECoG raw data by four experts.* (a) *Propagation of HFOs.* (b) *Propagation of rhythmic $\vartheta$-activity.*

# 5 HFO Detection

> Causa latet, vis est notissima fontis.
>
> — Publius Ovidius Naso: *Metamorphoses*
>
> The cause is hidden. The effect is visible to all.
>
> — *Metamorphoses*[1] by Ovid (43 BC - 17), Roman poet

This chapter is based on material which has already been published together with co-workers. We refer to Graef, Flamm, Pirker, Baumgartner, Deistler and Matz (2013) for the original article.

## 5.1 Introduction

### 5.1.1 Background

In this chapter we are concerned with the automatic detection of *high-frequency oscillations (HFOs)*. The popularity of these low-amplitude and high-frequency biomarkers lies in their excellent correlation with the SOZ, compare Subsection 2.3.5.

We employ a methodology initially proposed by Graef, Flamm, Pirker, Baumgartner, Deistler and Matz (2013) for the automatic detection of ictal HFOs in order to determine the initial seizure spread. This method is based on matched subspace filtering, compare Section 3.5 for an introduction to signal detection theory.

The automatic detection of HFOs based on classical signal detection methodology has received growing attention in the last years. We refer to

---

[1] Book 3; English translation by Anthony S. Kline.
Latin text accessible online at www.intratext.com/IXT/LAT0537.

| Publication | Data | Band-pass | Statistics | Threshold | Supp. |
|---|---|---|---|---|---|
| Staba et al. (2002) | Conventional depth electrodes; interictal | 100-500 Hz | RMS | mean + 5 SD | 6 ms |
| Smart et al. (2005) | Intracranial electrodes | 60-100 Hz | Teager energy | ? | ? |
| Nelson et al. (2006) | Micowire depth electrodes; interictal + ictal (animal) | ? | Teager energy | ? | ? |
| Gardner et al. (2007) | Conventional depth + subdural electrodes; interictal | 30-85 Hz (subsequent pre-emp. step) | Line length | 97.5% per-centile | 80 ms |
| Worrell et al. (2008) | Conventional + microwire depth electrodes; interictal | 80-1000 Hz (subsequent pre-emp. step) | Line length | 95.0% per-centile | 80 ms |
| Crépon et al. (2010) | Conventional depth + subdural electrodes; interictal | 180-400 Hz | Envelope via Hilbert transform | mean + 5 SD | ? |

**Table 5.1.1:** *Overview of basic HFO detection approaches.* Subsequent to band-pass filtering in the frequency band of interest statistics measuring the power contents in this band are considered. Continuous threshold exceedings during a suppression period result in an HFO detection.

the master thesis by Chander (2007) for an introduction to computational HFO detection approaches.

The methods initially proposed are rather simple and straight-forward in nature. They all follow the same approach, compare the overview in Table 5.1.1.

The common idea is the following: Subsequent to band-pass filtering in the frequency band of interest (e.g. ripples, fast ripples) one tries to measure the power content in this band. For this purpose, various statistics have been used, e.g. *RMS*[2] (Staba et al. 2002), *Teager energy*[3] (Smart et al. 2005), *Line length*[4] (Gardner et al. 2007) or the signal envelope based on the Hilbert transform (Crépon et al. 2010). An HFO detection is registered whenever such a statistic continuously exceeds a threshold (derived from a reference period) during a suppression period. While this duration is mostly manually imposed, its automatic determination supports a stable measurement of the HFO rate (Zelmann et al. 2009).

Additionally, in order to compensate for the spectral roll-off of EEG data, pre-emphasis steps have been proposed in some of these approaches, e.g. pre-whitening in Worrell et al. (2008). These methods have been applied to such different data as interictal and ictal recordings from microwire depth electrodes and conventional intracranial electrodes (depth as well as subdural electrodes), compare Table 5.1.1. Furthermore, wavelet analysis of evoked potentials in the HFO frequency range has been performed (van 't Klooster et al. 2011).

As the simple approaches listed in Table 5.1.1 are sensitive, but not highly specific, increasingly complicated multi-step approaches have been proposed in order to avoid false-positive detections. Very recently a number of them has been published, compare e.g. Zelmann et al. (2012) based on Zelmann et al. (2010), Blanco et al. (2010), Blanco et al. (2011) and von Ellenrieder et al. (2012).

## 5.1.2 Contribution

The visual inspection of ECoG data for HFO analysis is a time-demanding, highly subjective task which depends heavily on the individual experience of the investigator. Therefore, we pursue a complementary computational approach based on classical signal detection methodology initially proposed

---

[2]The *root mean square* (abbreviated RMS), also known as the *quadratic mean*, is a statistical measure of the magnitude of the (finite-energy) signal $x[n]$, $RMS = \sqrt{\sum_n x[n]^2}$.
[3]See Kaiser (1990) for its initial definition.
[4]See Esteller et al. (2001) for its initial definition.

in Graef, Flamm, Pirker, Baumgartner, Deistler and Matz (2013), compare Section 3.5.

While the studies mentioned in Subsection 5.1.1 focus on the automatic detection of HFOs in interictal invasive EEG (or in databases including both ictal and interictal phases), we limit ourselves to the analysis of ictal ECoG recordings. In contrast to the aforementioned approaches, we are not interested in the generation of statistics of HFO rates, but want to track the spread of initial HFOs. Channels showing first HFOs indicate the SOZ (of conventional ictal activity). For this analysis, seizure onset time, i.e onset of HFOs according to Section 4.3, is provided by clinicians.

Note that the visual inspection did not reveal any fast ripples, thus this study is limited to the ripple band.

## 5.2 Method

In this chapter, preprocessing of the ECoG data differs from the one detailed in Section 4.3, in particular from the downsampling step to 128 Hz. As we are interested in frequencies above the Nyquist frequency of 64 Hz, signals were low-pass filtered at 256 Hz in order to avoid aliasing and then only downsampled to 512 Hz sampling rate. All other preprocessing steps, e.g. notch filtering of line interference, are performed as listed in Section 4.3.

### 5.2.1 Matched subspace filter

We employ classical methods of signal detection to detect HFOs in the ripple band, compare Subsection 5.1.1. However, a major problem of the HFO detection approaches listed in Table 5.1.1 is the low specificity, i.e. these methods are likely to classify sharp transients as HFOs. We will discuss this phenomenon in Subsection 5.3.1.

In order to avoid this behavior, we make use of matched subspace filtering, namely an implementation of the CFAR (constant false alarm rate) matched subspace filter (3.5.4) according to Scharf (1991). The main idea of this detector is to compare the power of the signal content in the ripple band to the power of the residual signal, compare Subsection 3.5.3.

As a motivation, we start with a special case. Let $\mathbf{s} = \mathbf{H}\boldsymbol{\theta} \in \mathscr{S}$ be a linear combination of say $p$ pure sinuoidal signals in the ripple band (with a certain frequency resolution defined by $\Delta f = \text{floor}((250-80)/p)$),

$$\begin{aligned} s[n] &= \sum_{i=1}^{p} \theta_i \mathbf{h}_i[n] \\ &= \theta_1 \sin(2\pi n T\, 80) + \theta_2 \sin(2\pi n T\, (80+\Delta f)) + \\ &\quad + \theta_3 \sin(2\pi n T\, (80+2\Delta f)) + \ldots + \theta_p \sin(2\pi n T\, 250). \end{aligned}$$

In this case, the $N \times p$ modal matrix is

$$\mathbf{H} = \begin{pmatrix} \sin(2\pi T\, 80) & \cdots & \sin(2\pi T\, 250) \\ \vdots & \ddots & \vdots \\ \sin(2\pi N T\, 80) & \cdots & \sin(2\pi N T\, 250) \end{pmatrix},$$

and $\mathscr{S} \subset \mathbb{R}^N$ is the column space of $\mathbf{H}$, i.e. a hyper-plane. Furthermore, the projection matrix onto the subspace $\mathscr{S}$ is

$$\mathbf{P}_{\mathscr{S}} = \mathbf{H}(\mathbf{H}^T \mathbf{H})^{-1} \mathbf{H}^T, \tag{5.2.1}$$

and the quadratic form $\mathbf{x}^T \mathbf{P}_S \mathbf{x}$ in (3.5.4) measures the power contribution of the part of the signal corresponding to the subspace $\mathscr{S}$.

Although this modeling approach is easy to interpret[5], it has major disadvantages. First, the linear combination of $p$ sinusoidal signals only allows to capture activity at a finite number of sharp frequencies, not of the whole frequency band. However, neuronal activity is unlikely to be concentrated at constant, isolated frequencies, but rather is smeared in a whole frequency band. Second, if we aim at an increase of the frequency resolution in the regression, we have to decrease $\Delta f$. However, this blows up the dimension of the modal matrix $\mathbf{H}$ and leads to numerical problems of the inversion in (5.2.1) due to the bad conditioning of $\mathbf{H}$.

---

[5] As in Subsection 5.1.1, we have a linear combination of modes (the sinusoidal signals), and $\|\boldsymbol{\theta}\| > 0$ indicates a contribution of $\mathbf{s}$ to $\mathbf{x}$, i.e. the presence of ripples.

Thus, in order to obtain the signal contribution in the ripple band, we prefer to filter the ECoG data rather than to regress them on sinusoidal signals. For this purpose, we proceed as follows:

We consider an idealized band-pass filter with transfer function

$$H(f) = \begin{cases} 1 & f \in [80, 250] \\ 0 & f \notin [80, 250] \end{cases}. \tag{5.2.2}$$

Now, $\mathscr{S}$ is the signal subspace (of infinite dimension) of stationary signals band-limited to the ripple band $[80, 250]$, as indicated by (5.2.2). Consequently, we have an infinite weighting sequence $\theta_i, i \in \mathbb{Z}$. The filtering operation (5.2.2) is the orthogonal projection operator onto $\mathscr{S}$, as can be seen immediately from the spectral representation of a stationary process (Brockwell and Davis 1991).

Thus, we consider the projection operator $P_\mathscr{S} = H(f)$ in this setting, and the quadratic form $\mathbf{x}^T P_\mathscr{S} \mathbf{x}$ in (3.5.4) measures the power contribution of the part of the signal corresponding to the signal subspace $\mathscr{S}$. Note that we now consider the power contribution of the entire ripple band, not just activity at isolated frequencies.

Therefore, we proceed as follows for each channel $x_k[n]$, $k = 1, \ldots, N$:

We initially perform a pre-emphasis step. We apply a high-pass filter at 13 Hz (upper bound of the $\alpha$-band) to the signal $x_k[n]$ in order to compensate for the strong spectral roll-off of ECoG data. We denote the spectrally equalized signal by $\tilde{x}_k[n]$. The choice of this high-pass filter will be discussed in Subsection 5.3.1.

Second, we apply a band-pass filter in the frequency range of 75-250 Hz to obtain $\tilde{x}_k^R[n]$. This extended band allows to reliably capture HFOs with a frequency at the lower end of the ripple band (80 Hz or slightly below).

Third, we calculate the power ratio for the CFAR matched subspace filter (3.5.4). In this application, we follow a common approach in engineering (Randall and Tech 1987), in particular in HFO analysis (Crépon et al. 2010), for determining the power in a frequency band by means of Hilbert transformation. We calculate the statistic (3.5.4) as

$$T^2[n] = \frac{\left|\mathcal{H}\{\tilde{x}_k^R\}[n]\right|^2}{\left|\mathcal{H}\{\tilde{x}_k - \tilde{x}_k^R\}[n]\right|^2}, \qquad (5.2.3)$$

where $\mathcal{H}\{x\}[n] = x[n] + i\check{x}[n]$ denotes the (complex-valued) analytic signal, which is obtained by Hilbert transformation $\check{x}$ of the signal $x$ at time-point $n$ (Gabor 1946). Note that we consider a definition of the Hilbert transform for stationary signals according to Lindgren (2012), rather than the classic one for finite-energy signals:[6] Let $\mathcal{H}\{x\}[n]$ be half of the spectral representation of a stationary process, i.e. twice the spectral representation integral for positive frequencies. Then the Hilbert transform of the stationary process $(x[n], n \in \mathbb{Z})$ is the process $(\check{x}[n], n \in \mathbb{Z})$ in $\mathcal{H}\{x\}[n] = x[n] + i\check{x}[n]$ and is the result of a linear filter on $x[n]$ with frequency response $H(f) = -\text{sgn}(f)i$.[7] As is well known (Ville 1948), the absolute value of the (complex) analytic signal, $|\mathcal{H}\{x\}[n]|$, represents the envelope of $x$ at time-point $n$, and the squared envelope, $|\mathcal{H}\{x\}[n]|^2$, indicates the power of the signal $x$. Thus, the numerator in (5.2.3) represents the power in the ripple band, the denominator the power of the residual signal.

Finally, we detect the presence of HFOs if $T[n] > \gamma$. The threshold $\gamma$ is determined from a reference period prior to the ictal activity. In order to suppress false-positives due to sharp transients, we demand $T[n] > \gamma$ continuously for at least 50 ms, which corresponds to 4 cycles of an 80 Hz oscillation. This number of consecutive oscillations is often required in HFO literature, see e.g. Zijlmans et al. (2011).

---

[6] In signal processing, the Hilbert transform is commonly defined for finite-energy functions $f(t)$ as the Cauchy principal value of the convolution of $f$ with $1/(\pi t)$, as first analyzed by Hilbert (1912). In this case, the assumption of $f \in \mathscr{L}^p, 1 \leq p < \infty$, assures the existence of the convolution integral, compare Riesz (1928) and Titchmarsh (1962).
[7] Note that this definition preserves the properties of the classic definition of the Hilbert transform for finite-energy signals, e.g. phase shift of $\pi/2$ due to the design of the frequency response and representation of the positive half of the spectrum by the analytic signal.

## 5.2.2 Signal model

In order to demonstrate the effectiveness of the proposed algorithm, we test it on simulated data in Subsection 5.3.1.

For this purpose we fit an AR-8 model to channel A12 in a 20-second lasting period 30 seconds prior to seizure 1. Based on these parameter estimates, we simulate 5 seconds of an AR-8 signal $s[n]$.[8] The test signal then contains superposed HFOs during 0.5 seconds, i.e.

$$x[n] = \begin{cases} s[n] & n = 1\ldots 2f_s, \\ s[n] + a\sin\left(\frac{2\pi n f}{f_s}\right) & n = 2f_s + 1\ldots 2.5f_s, \\ s[n] & n = 2.5f_s + 1\ldots 5f_s. \end{cases} \quad (5.2.4)$$

The variance of the white noise in the AR simulation step was set to $\sigma^2 = 5$, which resulted in a small HFO-to-background SNR of $-2.7$ dB.

In order to facilitate the comparison with ECoG data, a sampling frequency $f_s$ of 512 Hz is used for simulation. We set the frequency to $f = 85$ and the amplitude to $a = 12$ to obtain simulated ripples, compare plot (a) of Fig. 5.3.1.

## 5.3 Results

### 5.3.1 Signal model

In order to assess our methodology, we apply the HFO detection algorithm to the test signal (5.2.4). In this artificial setting, the threshold $\gamma$ is calculated as the 90%-percentile of $T[n]$ from the simulated AR signal $s[n]$.

Fig. 5.3.1 details the results: The simulated HFOs are correctly detected in the time interval 2.0-2.5 s, see the HFO detection sequence in plot (d).

---

[8] AR estimation and subsequent simulation of this model were done with the help of the Matlab® package *arfit* (methods *arfit* and *arsim*) developed by Schneider and Neumaier (2001).

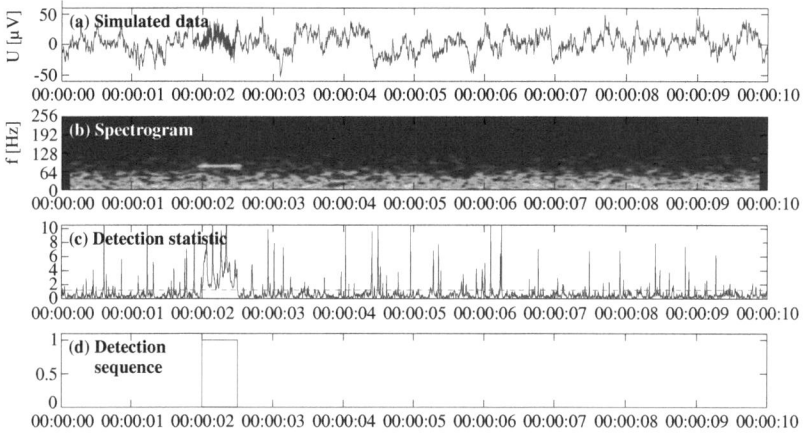

**Figure 5.3.1:** *HFO detection: simulation results. (a) Simulated test signal, (b) spectrogram, (c) statistic $T[n]$ and threshold $\gamma$, (d) HFO detection sequence. Simulated HFOs are detected.*

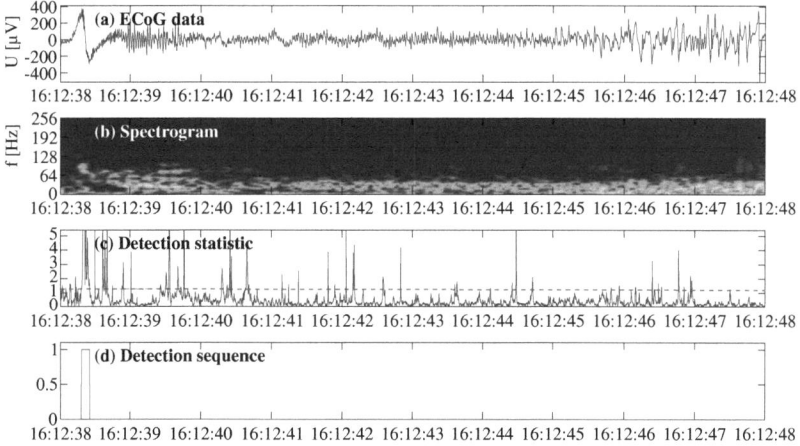

**Figure 5.3.2:** *HFO detection: results for seizure 1, initial 10 seconds of channel B8. (a) ECoG data, (b) spectrogram, (c) statistic $T[n]$ and threshold $\gamma$, (d) HFO detection sequence. HFOs are detected.*

During this period the statistic $T[n]$ takes large values, as can be seen from plot (c). Sporadic short threshold exceedings of $T[n]$ are successfully suppressed (avoiding false-positives).

Note that for better visualization a spectrogram is shown in plot (b). It clearly reveals the presence of the simulated HFOs in the time interval 2.0-2.5 s.

### 5.3.2 HFO propagation

We apply the proposed methodology to the first two seizures. In each case, the threshold $\gamma$ was determined as the 90%-percentile of the statistic $T[n]$ from a 120-second reference period starting three minutes prior to the respective seizure.

Let us first have a look at detailed detection results of channel B8 in seizure 1. For this purpose, we consider Fig. 5.3.2 which is built up in analogy to Fig. 5.3.1 from Subsection 5.3.1. Again, plot (a) shows the raw ECoG data and plot (b) the corresponding spectrogram. In the depicted time scale (10 s shown in Fig. 5.3.2), HFO activity is difficult to detect in the raw data by visual analysis. However, it can be easily recognized in the spectrogram around 16:12:38.500. As expected, the statistic $T[n]$ takes large values at that time, compare plot (c). Furthermore, sporadic short threshold exceedings of $T[n]$ throughout the ten seconds are successfully suppressed. This specific behavior will be subject to further discussion in Subsection 5.4.1.

In Fig. 5.3.3 we present the results for the HFO propagation analysis of the full channel set in both seizures. Detected HFOs, i.e. signals in periods with a non-zero HFO detection sequence in Fig. 5.3.2 (d), are highlighted in red. We distinguish between initial, close follow-up (within 250 ms) and later HFO activity. As a benchmark, three clinical experts independently analyzed the two seizures by visual inspection and marked the first occurrence of HFOs on each channel.

According to the algorithm, initial HFOs are found on channels B8 and B7 in case of seizure 1, see plot (a). A quick propagation (~ 125 ms) takes place to channels A11 and A5. Later HFOs are detected on electrodes A and C. In seizure 2 our algorithm detects initial HFOs on channel A9, compare

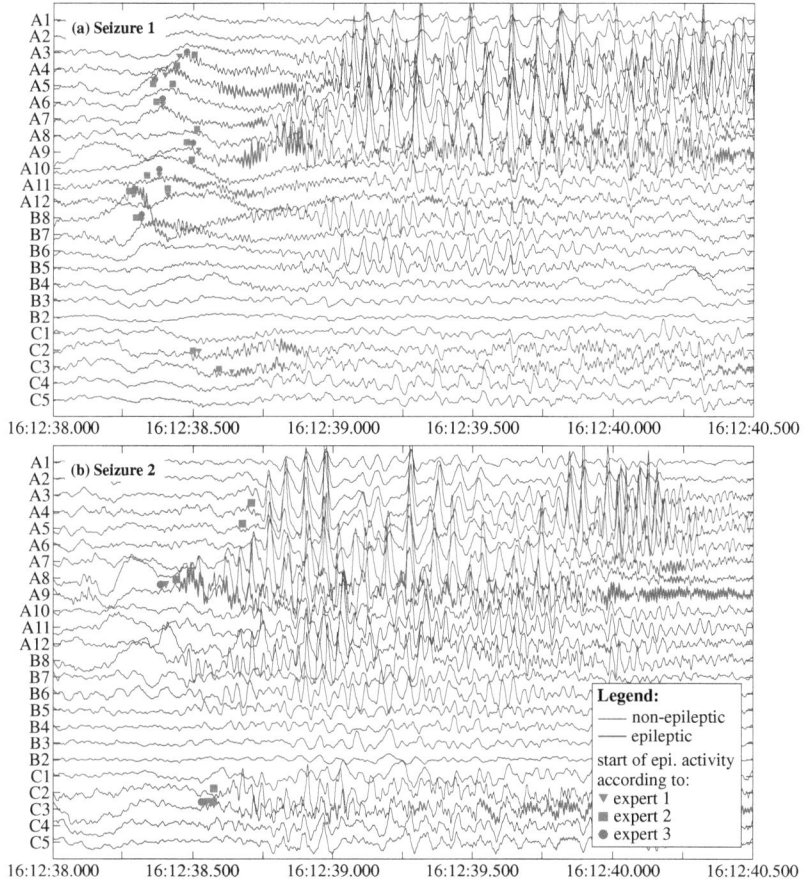

**Figure 5.3.3:** *HFO propagation.* *(a) Seizure 1, (b) seizure 2. Automatically detected HFO activity is highlighted in red. Onset of HFO activity according to the visual analysis of three clinical experts is indicated.*

plot (b). Here, we do not observe any close follow-up HFO activity on other channels. These findings correlate well with the visual analysis of the clinical experts, see Fig. 5.3.3.

Table 5.3.1 summarizes the findings of our algorithm and the visual inspection for initial and close follow-up activity. According to the automated analysis, the HFO generating zone comprises the parieto-occipital area

| Seizure | Examiner | Initial HFOs | | Conventional $\vartheta$-activity | |
|---|---|---|---|---|---|
| | | initial | follow-up | initial | follow-up |
| 1 | Algorithm | B8, B7 | A11, A5 | - | - |
| | Expert 1 | B8, B7 | A11, A5, A6, A7, A12 | B8 | A10, A11, A12 |
| | Expert 2 | B8, B7 | A11, A5, A6, A7, A12 | A11, A12, B8 | A9, A10, B7 |
| | Expert 3/4 | B8, B7 | A5, A7, A11 | A10, A11, A12 | B8 |
| 2 | Algorithm | A9 | - | - | - |
| | Expert 1 | A9 | C3 | A11, A12 | A9, A10 |
| | Expert 2 | A9 | C3, C2 | A11, A12 | A10 |
| | Expert 3/4 | A9 | C3 | A11, A12 | B8 |

Table 5.3.1: *HFO propagation. Results of the HFO detection algorithm in first two seizures and comparison to visual inspection of HFO- and conventional $\vartheta$-activity.*

between electrodes B7 and A9 (compare Fig. 5.4.3 in Subsection 5.4.2). The correlation between the onset of HFOs and $\vartheta$-activity (listed in the column on the right) will be discussed in Subsection 5.4.2.

## 5.4 Discussion

### 5.4.1 Matched subspace filter

In this subsection we briefly want to discuss two details of the HFO detection algorithm outlined in Subsection 5.2.1, the matched subspace filtering approach and the pre-emphasis step.

First, an advantage of the proposed algorithm based on matched subspace filtering lies in its reduction of false-positive detections.

As mentioned in Subsection 5.1.1, the simple approaches listed in Table 5.1.1 have the disadvantage of a low specificity due to false-positive detections in case of sharp transients. The reason for this behavior lies in the band-pass filtering step: Very sharp transients are similar in morphology to Dirac impulses, and the impulse response of the FIR band-pass filter is the set of filter coefficients (Oppenheim and Schafer 1989). Consequently, even in case of little or no energy in the ripple band of the raw ECoG data, the band-pass-filtered signal contains energy in the entire frequency band of the FIR filter. Thus, the statistic $T[n]$ behaves as if HFOs were present.

Consider Fig. 5.4.1 for an illustrative example. It details the results for an implementation of the RMS-based algorithm proposed by Staba et al. (2002).[9] As in Subsection 5.3.2, plot (a) depicts 10 seconds of ECoG data. Note the very sharp transients in the last two seconds (16:12:46 to 16:12:48). These transients provoke considerable threshold exceedings, which last for longer than the imposed suppression duration, compare plot (c). This results in numerous false-positive HFO detections, compare the detection sequence in plot (d).

These band-pass filtering artifacts motivate the use of matched subspace filtering: By normalizing the power in the ripple band to the power of the residual signal one aims at diminishing the energy induced by filtering. Fig. 5.3.2 in Subsection 5.3.1 confirms this hypothesis: Threshold exceedings due to sharp transients are either completely suppressed or reduced in duration, compare plot (c). In the latter case these short exceedings are successfully suppressed by the imposed suppression duration of 50 ms, compare plot (d).

Second, we want to provide a justification for the design of the high-pass filter in the pre-emphasis step. As mentioned in Subsection 5.2.1, the strong spectral roll-off of EEG data requires a pre-emphasis step. A spectral equilibration is often performed by a high-pass filter with a frequency response smoothly increasing over the entire frequency range: For instance, in EEG analysis this may be achieved by pre-whitening (Worrell et al. 2008), in speech processing one typically designs high-pass filters with frequency responses proportional to the inverse of the spectrum (Picone 1993).

In contrast, we choose a high-pass filter which strictly eliminates frequency contributions below 13 Hz and lets pass all others. The advantage of this approach lies in filtering out strong-power contributions in the physiological frequency bands ($\delta$, $\vartheta$, $\alpha$) we are not interested in in this context.

---

[9]Compare Table 5.1.1 for the implementation details.

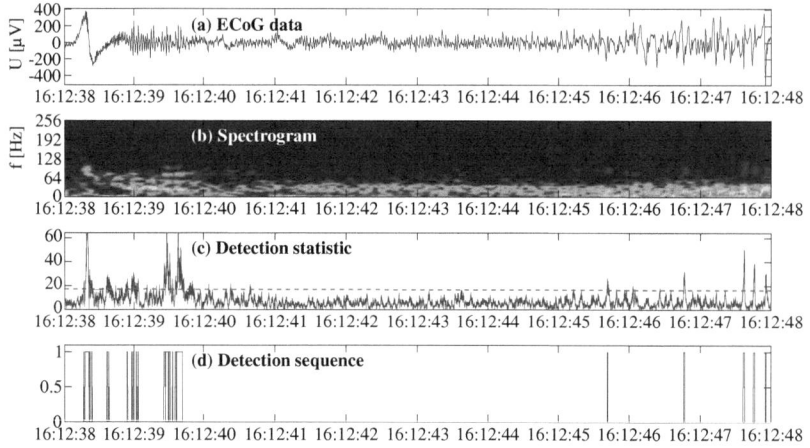

**Figure 5.4.1:** *HFO detection via RMS in seizure 1*, initial 10 seconds of channel B8. (a) ECoG data, (b) spectrogram, (c) statistic $T[n]$ and threshold $\gamma$, (d) HFO detection sequence. HFOs are detected, but false-positives appear due to sharp transients.

Therefore, these low-frequency bands do not influence ratio (5.2.3), and statistic $T[n]$ measures low-power HFO activity with higher precision.

### 5.4.2 HFO propagation

In Subsection 5.3.1 we showed that our proposed methodology is capable of detecting HFOs in simulated data while suppressing sporadic false alarms. The application of this methodology to invasive EEG recordings (Subsection 5.3.2) yields promising results as well.

In both seizures our method correctly identifies the electrodes with initial HFO activity, B7 and B8 in seizure 1, A9 in seizure 2 (see Table 5.3.1). These findings match the visual analysis of all three experts.

In case of follow-up HFO activity, our algorithm also yields good results: In seizure 1 we successfully mark close HFO follow-up activity on electrodes A11 and A5. Later HFO activity on other electrodes (A7, A9, C2, C3) is correctly identified, but detection is delayed up to 250 ms, compare Fig. 4.3.2. In seizure 2 the experts flag electrodes C2 and C3 as

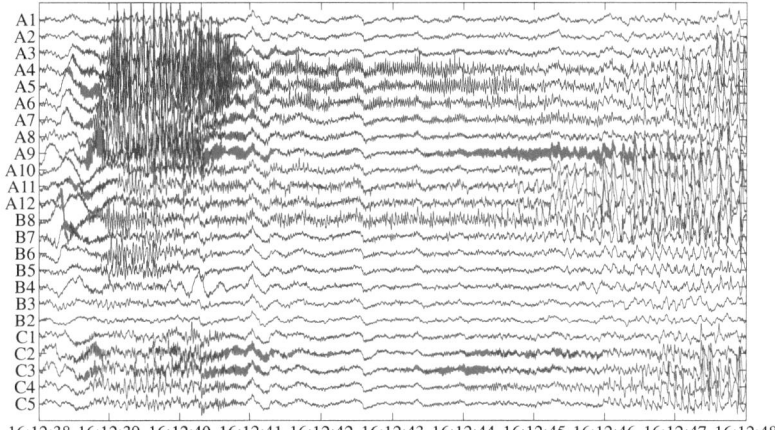

**Figure 5.4.2:** *HFO propagation in seizure 1, initial 20 seconds. Area of initial ictal HFOs and onset area of conventional $\vartheta$-activity are well correlated (B8, A12).*

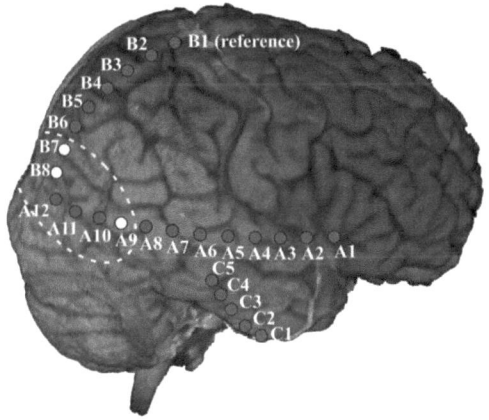

**Figure 5.4.3:** *SOZ according to HFO detection on MRI scan with electrodes positions. Electrodes revealing initial ictal HFO activity are marked in white (seizure 1: B7, B8; seizure 2: A9). The supposed seizure onset zone is outlined.*

follow-up, whereas our algorithm marks late HFO propagation on C3 with a latency of 500 ms.

The area of initial HFO activity (B7, B8 and A9) correlates well with the SOZ determined by visual inspection of conventional $\vartheta$-activity, compare Table 5.3.1. Fig 5.4.2 shows that the time interval between the occurrence of ictal HFOs and the onset of conventional $\vartheta$-activity is 7 s (seizure 1) and 8 s (seizure 2), which is in good accordance with literature (compare Subsection 2.3.5). Note the the long-lasting HFO activity on channel A9, which is part of the SOZ. We speculate that such prolonged ictal HFO activity could provide evidence for localizing the SOZ.[10]

Based on the results of our HFO analysis we infer that the SOZ comprises the parieto-occipital area between electrodes B7 and A9, see Fig. 5.4.3 for a visualization.

### 5.4.3 Concluding remarks

In this chapter we proposed a novel method based on matched subspace filtering for the detection of HFOs in the ripple band. This pilot study shows promising first results in tracking of ictal HFO propagation as an indicator for the SOZ. Therefore, we are confident that our method has the potential for an objectivation in the presurgical clinical examination of therapy-resistant patients.

Next necessary steps include the application to a broader data basis and a subsequent statistical analysis (e.g. detection rate/false positives, mean latency in detection) for better understanding of the performance of our method.

---

[10] To our knowledge, this potential correlation has not been examined yet.

# 6 Causality Analysis

> Felix qui potuit rerum cognoscere causas.
>
> — Publius Vergilius Maro: *Georgica*
>
> He who's been able to learn the causes of things is happy.
>
> — *Georgics*[1] by Virgil (70 - 19 BC), Roman poet

This chapter is based on material which has already been published together with co-workers. We refer to Graef, Hartmann, Flamm, Baumgartner, Deistler and Kluge (2013) for the original article.

## 6.1 Introduction

### 6.1.1 Background

In this chapter we want to visualize coupling effects of the multivariate ECoG signal in order to localize the SOZ. For this purpose, we calculate dependency measures and depict them in a graph, whose vertices represent the components of the signal, and edges indicate dependencies different from zero. This approach yields an intuitive graphical representation of coupling effects in multivariate signals (Dahlhaus 2000, Dahlhaus and Eichler 2003, Eichler 2006a).

A wide variety of different dependency measures for neurophysiological data has been published, see Section 3.4. Two important directed coupling indicators are the *Directed Transfer Function (DTF)* and the *Partial Directed Coherence (PDC)*, both distinguishing between source and target by indicating a direction of the dependency. DTF and PDC are based on a

---
[1] Book 4, verse 290; English translation by A. S. Kline.
Latin text accessible online at virgil.org.

common linear approach to EEG analysis, i.e. multivariate autoregressive modeling, compare Section 3.2. In case of DTF and PDC, the parametric estimation of the spectrum is the basis for the measurement of linear couplings in the frequency domain, which reveal relations between electrodes. These inter-dependencies are interpreted as indications for epileptic synchronous activity.

DTF was proposed by Kaminski and Blinowska (1991) for quantifying dependencies in neural signals (see Subsection 3.4.3). However, when Franaszczuk et al. (1994), Franaszczuk and Bergery (1998) and Ge et al. (2007) used DTF for epileptic seizure analysis, a manual selection of narrow frequency bands always had to be performed in order to achieve satisfying results. Wilke et al. (2008) proposed a time-variant version of DTF for epileptic EEG analysis, which shows promising first results. Recently Kim et al. (2010) combined the usage of DTF with a spatio-temporal source localization algorithm in order to analyze the propagation of epileptic activity in ECoG signals.

PDC was proposed by Baccala and Sameshima (2001) and has been receiving growing attention since then, including extensions and theoretical considerations of its properties (see Subsection 3.4.4). It has often been applied to the analysis of neural interactions, compare e.g. Sameshima and Baccala (1999) and Astolfi, Cincotti, Babiloni, Carducci, Basilisco, Rossini, Salinari, Mattia, Cerutti, Dayan, Ding, Ni, He and Babiloni (2005).

## 6.1.2 Contribution

Our aim is to analyze synchronization effects in multichannel ECoG data of epileptic patients and to identify these coupling effects with a high degree of automation: Unlike the methods mentioned above, we neither want to manually preselect ECoG input channels nor explicitly consider specific frequency bands. For this purpose we make use of a methodology initially proposed in Graef, Hartmann, Flamm, Baumgartner, Deistler and Kluge (2013):

In order to avoid numerical problems in the AR model estimation due to the high number of ECoG channels, we employ an automatic channel selection procedure prior to computing dependency measures. This idea is detailed in Subsection 6.2.3.

Furthermore we follow a novel approach to the identification of synchronous activity: Contrary to a spectral analysis, as it is performed by DTF or PDC, we consider a time domain approach. In order to assure a (neuro)physiological interpretation of our methodology, we search for a coupling indicator with a clear physical interpretability. For this purpose we use a dependency measure termed *EIPR (extrinsic-to-intrinsic-power-ratio)* initially defined by Hartmann et al. (2008), which is discussed in Subsection 6.2.4.

## 6.2 Method

### 6.2.1 Autoregressive model

Our methodology is based on autoregressive modeling, compare Section 3.2. The AR(p) model (3.2.1) is decomposed component-wise into the separated contributions of all channels: We define the *partial contribution* $\mu_{k,l}[n]$ as

$$\mu_{k,l}[n] \triangleq \sum_{s=1}^{p} A_{k,l}[s] x_l[n-s] \qquad (6.2.1)$$

with $A_{k,l}[s]$ the (k,l)-element of the coefficient matrix $\mathbf{A}[s]$ in (3.2.1).[2] This allows to write the AR(p) model (3.2.1) for each channel $x_k[n]$, $k = 1, \ldots, K$ as

$$x_k[n] = \mu_{k,k}[n] + \sum_{l \neq k} \mu_{k,l}[n] + \varepsilon_k[n].$$

In order to shrink the regression model, we only consider partial contribution terms $\mu_{k,l}[n]$ which significantly differ from zero. The explicit choice of

---
[2] A more general definition of the partial contribution term (6.2.1), as proposed by Hartmann et al. (2008), would allow a more flexible lag usage, e.g. permitting non-causal modeling.

regressors yields a model of the form

$$x_k[n] = \mu_{k,k}[n] + \sum_{l \in \mathbb{L}_k} \mu_{k,l}[n] + \tilde{\varepsilon}_k[n], \quad k = 1,..,K. \qquad (6.2.2)$$

Here, $\mathbb{L}_k$ is an *extrinsic channel set*, which can be a subset of the set of all extrinsic channels, $\{1,\ldots,K\}\setminus\{k\}$, allowing for a reduction of the number of parameters of the AR model. A strategy for such a reduction is proposed in Subsection 6.2.3.

Thus, each $\mu_{k,l}[n]$ in equation (6.2.2) reflects the contribution from (the past of) the respective channel $x_l[n]$ to channel $x_k[n]$. As we differentiate between the channel $x_k[n]$ and the other $x_l[n]$, $l \neq k$, in equation (6.2.2), we introduce the following specification: For $k = l$, we call $\mu_{k,k}[n]$ the *intrinsic contribution*; for $k \neq l$, $\mu_{k,l}[n]$ is the *partial extrinsic contribution*. As the term $\sum_{l \in \mathbb{L}_k} \mu_{k,l}[n]$ in equation (6.2.2) is the sum of all partial extrinsic contributions, it symbolizes the total amount of inflow to the channel $x_k[n]$ and is therefore denoted by *total extrinsic contribution*.

### 6.2.2 Solution of the normal equations

Under the assumption of stationarity, the solution of the ordinary-least-squares (OLS) normal equations within the data window yields the estimated model coefficients $A_{k,l}[s]$, compare Graef (2008). Second-order-statistics needed for their solution have to be estimated from the data. For this reason it is important that the length of the data window is chosen neither too short nor too long. An appropriate choice has to establish a good trade-off between estimation errors due to instationarity (bias) and inaccuracy due to a too small number of samples (variance).

In particular in case of neural data such as ECoG recordings their apparently highly instationary character (compare Figs. 4.3.2, 4.3.3 and 4.3.4) requires the use of short data windows.

### 6.2.3 Dynamic input channel selection

The estimation of the model coefficients $A_{k,l}[s]$ in the normal equations poses numerical problems, as we deal with a large number of ECoG channels which are highly correlated both in time as well as in the cross-sectional dimension. In order to avoid this situation, the idea is therefore to automatically reduce the number of channels in a subset containing all information important for the regression.

For this reason we introduced the extrinsic channel set $\mathbb{L}_k$ in equation (6.2.2), which defines – per channel $x_k[n]$ – the $x_l[n]$ relevant for the autoregressive model. The advantage arising from this approach is that we do not have to choose $\mathbb{L}_k = \{1,\ldots,K\}\setminus\{k\}$ (as it would be the case in the AR model (3.2.1)), but can shrink it to a reduced set of channels. We only consider the channels $\{x_k, x_l : l \in \mathbb{L}_k\}$, and this selection assures that the estimation of the model coefficients yields numerically stable results: The correlation matrix of the small subsystem is well conditioned and can be inverted without further numerical problems.

We propose an iterative procedure for an automatic selection of an extrinsic channel set $\mathbb{L}_k$ for each $x_k$, which is described in pseudo code in Fig. 6.2.1.

The main idea is to iteratively add channels in a bottom-up fashion until an information criterion is minimized. Here we make use of the well-known *Akaike information criterion (AIC)*[3] which is defined in this context as (Penm and Terrell 1982)

$$AIC(\mathbb{L}_k) \triangleq \ln S_{\text{err}}(\mathbb{L}_k \cup \{k\}) + \frac{2M}{N_{win}}, \qquad (6.2.3)$$

where

$$S_{\text{err}} \triangleq \sum_{n=1}^{N_{win}} (\tilde{\varepsilon}_k[n])^2$$

---

[3] Note that in Graef, Hartmann, Flamm, Baumgartner, Deistler and Kluge (2013) we use the Bayesian information criterion (BIC) introduced by Schwarz (1978).

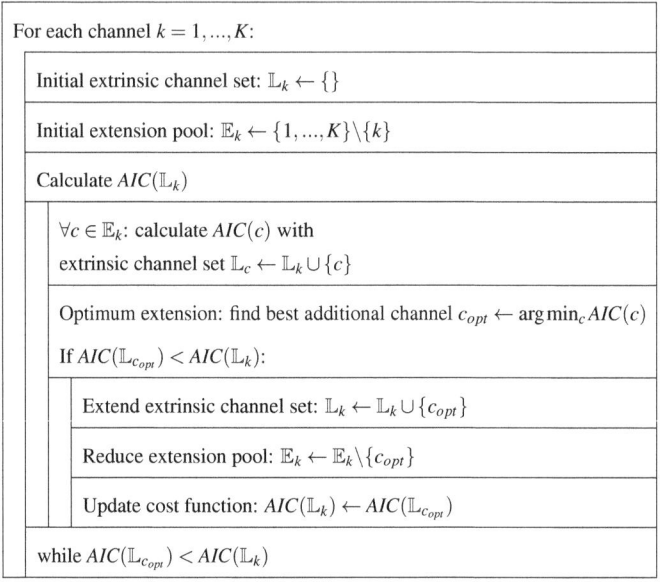

**Figure 6.2.1:** *Automatic channel selection algorithm.* The bottom-up construction of the extrinsic channel set $\mathbb{L}_k$ is given in pseudo code.

is the residual sum of squares and

$$M \triangleq \left( \sum_{s=1}^{K} \delta_s(\mathbb{L}_k) + 1 \right) \cdot p$$

with

$$\delta_s(\mathbb{L}_k) \triangleq \begin{cases} 1 & s \in \mathbb{L}_k \\ 0 & s \notin \mathbb{L}_k \end{cases}.$$

Hence $M = (\dim \mathbb{L}_k + 1) \cdot p$ is the total number of parameters to be estimated.

Using this criterion our algorithm works as follows: We start with an empty extrinsic channel set $\mathbb{L}_k$. Then we add the channel $x_l[n]$, $l \neq k$ of the $(K-1)$ other ones which is best in the sense that it leads to the smallest AIC value (6.2.3). In the next step we again select the »best« out of the remaining ones and so forth till we cannot decrease the value of expression (6.2.3) any more by adding channels. This (local) minimum determines the extrinsic channel set $\mathbb{L}_k$ to be used for coefficient estimation. Coefficients $A_{k,l}[s]$ of channels $x_l[n]$ which were not selected by this iterative procedure are set to zero.

We expect the algorithm to select extrinsic channels which contribute significantly to the explanation of the respective intrinsic channel. We will illustrate this behavior in Subsection 6.3.2 in detail.

Note that the bottom-up approach of our proposed algorithm is similar to the *An algorithm* published by An and Gu (1989), which is however limited to a regression model without any temporal lags.

### 6.2.4 Partial extrinsic power

The goal of the proposed method is to identify directed dependencies of the multivariate signal $\mathbf{x}[n]$, which are expected to indicate synchronization and coupling effects of brain regions during epileptic seizures. For similar problems, numerous alternative measures based on a spectral analysis have been proposed in literature, as mentioned in Subsection 6.1.1. However, instead of regarding spectral properties of the AR model (6.2.2), we propose to directly consider the partial contribution term (6.2.1) in order to gain information on the influence of channel $x_l[n]$ to channel $x_k[n]$:

The variance of the partial contribution term $\mu_{k,l}[n]$ can be written as[4], using (6.2.1),

$$\mathbb{V}\{\mu_{k,l}[n]\} = \mathbb{E}\{\mu_{k,l}[n]\,\mu_{k,l}[n]\}$$
$$= \sum_{s=1}^{P}\sum_{s'=1}^{P} A_{k,l}[s]\, r_{x_l}[s-s']\, A_{k,l}[s'], \qquad (6.2.4)$$

where $r_{x_l}[s] = \mathbb{E}\{x_l[n+s]\, x_l[n]\}$ is the autocorrelation function of channel $x_l[n]$.

The partial contribution term $\mu_{k,l}[n]$ represents the directed influence of channel $x_l$ onto $x_k$ by construction, see model (6.2.2). Its variance is a natural measure of the strength of the influence from $x_l$ onto $x_k$. For $k = l$,

---
[4]Compare Graef (2008) for a detailed derivation.

we speak about the *intrinsic power*, for $k \neq l$ about the *partial extrinsic power*.

Note that although the sum of all partial extrinsic contributions $\mu_{k,l}[n]$ gives the total extrinsic contribution (i.e. the inflow from all channels $x_l[n]$ onto channel $x_k[n]$, $k \neq l$), the sum of all partial extrinsic power terms $\mathbb{V}\{\mu_{k,l}\}$ does not equal the total extrinsic power $\mathbb{V}\{\sum_{l \in \mathbb{L}_k} \mu_{k,l}[n]\}$, unless all cross-correlations are zero.

Considering the variance (6.2.4) of the respective partial contribution term for all channel combinations $k, l = 1, ..., K$, we expect to obtain an indication for the directed coupling of each channel $x_l[n]$ onto each channel $x_k[n]$ in scalar form.

## 6.2.5 Extrinsic-to-intrinsic-power-ratio (EIPR)

A problem with the variance (6.2.4) of the partial contribution term is its scale-dependence. It is desirable to normalize this measure appropriately such that it is independent of the signal power.

It is not obvious how to perform this normalization. One could, for example, normalize with respect to all target channels, as done by the Partial Directed Coherence (PDC) defined in (3.4.15). However, this approach renders the measure dependent of all channels involved in the regression (Schelter et al. 2009). This will be detailed in the next subsection.

This motivates our search for an alternative normalization which is not affected by this kind of limitation. We use the *extrinsic-to-intrinsic-power-ratio (EIPR)*

$$\eta_{k,l}^2 \triangleq \frac{\mathbb{V}\{\mu_{k,l}[n]\}}{\mathbb{V}\{\mu_{k,k}[n]\}}, \qquad (6.2.5)$$

which was initially defined by Hartmann et al. (2008). We assume that the variance of the intrinsic contribution term in the denominator in (6.2.5) is bounded below by a positive constant[5]. This assumption was justified in a

---

[5]The case of $\mathbb{V}\{\mu_{k,k}[n]\} = 0$ would imply that the past of $x_k$ does not contribute to the explanation of $x_k$ in the present, i.e. the chosen $x_l$ (alone) explain $x_k$ optimally. This is highly unlikely in ECoG data, which was empirically shown in Graef (2008).

study of ECoG recordings (Graef 2008).

EIPR defined in this way quantifies (directed) coupling effects of channel pairs $(x_k, x_l)$, taking large values for large partial extrinsic variance and small intrinsic variance. This is the case when channel $x_l$ contributes significant information to the explanation of channel $x_k$. On the other hand, EIPR shows only small values for weak influence of $x_l$ to $x_k$.

### 6.2.6 Comparison of EIPR and PDC

As mentioned in Subsection 3.4.4, PDC provides a »frequency-domain picture for Granger causality descriptions« (Baccala and Sameshima 2001): in particular $\pi_{k,l}^2(f) = 0 \ \forall f$ is equivalent to the statement that channel $x_l[n]$ does *not* Granger-cause channel $x_k[n]$ (compare Subsection 3.3.2 for Granger analysis in the frequency domain).

One important difference between EIPR and PDC lies in their respective normalization. As discussed in Subsection 3.4.4, PDC is normalized with respect to all target channels which renders the measure dependent of all channels involved in the regression (Schelter et al. 2009). Normalizing with respect to all source channels rather than to all target channels, as proposed by Schelter et al. (2009), causes similar problems.

A particular situation where a normalization either to all source or all target channels involved may lead to misleading interpretations is as follows: Imagine any arbitrary three-dimensional autoregressive model reflecting the dependencies depicted by Fig. 6.2.2 (b). When we study $\pi_{1,2}^2(f)$ indicating the directed coupling between $x_2$ and $x_1$, PDC is influenced by $x_3$, as

$$\pi_{1,2}^2(f) = \frac{|A_{1,2}(f)|^2}{|A_{1,2}(f)|^2 + |A_{2,2}(f)|^2 + |A_{3,2}(f)|^2}.$$

Thus, in case of ECoG signals, our observation of brain activity between two examined electrodes of interest is influenced by the measurement of a third electrode, which depend on its exact position on the cortex. Therefore, the analysis is impaired by the position of the third electrode, which we cannot adapt to our needs (as it is implanted).

EIPR avoids this problem, as its denominator is only based on the statistics of the intrinsic (currently regarded) channel (Graef et al. 2009).

Furthermore, it is interesting to note that the variance (6.2.4) of the partial contribution term is closely linked to PDC. Let us write this variance (6.2.4) as integral,

$$\mathbb{V}\{\mu_{k,l}[n]\} = \int_f S_{\mu_{k,l}}(f)\,df, \quad (6.2.6)$$

where $S_{\mu_{k,l}}(f)$ denotes the spectral density of the partial contribution term $\mu_{k,l}[n]$. By transforming the partial contribution term (6.2.1) into the frequency domain, i.e.

$$M_{k,l}(f) = A_{k,l}(f) X_l(f),$$

we obtain its spectral density

$$S_{\mu_{k,l}}(f) = |A_{k,l}(f)|^2 S_{x_l}(f). \quad (6.2.7)$$

Substituting expression (6.2.7) into the spectral representation (6.2.6), we obtain the representation

$$\mathbb{V}\{\mu_{k,l}[n]\} = \int_f |\tilde{A}_{k,l}(f)|^2 S_{x_l}(f)\,df, \ k \neq l \quad (6.2.8)$$

of the variance of the partial contribution term $\mu_{k,l}[n]$ (see (6.2.4) for its definition).

Hence, under the assumption that $S_{x_l}(f) > 0$ the left-hand side of (6.2.8) is zero if and only if PDC $\pi^2_{k,l}(f) = 0$, $k \neq l$ for all frequencies $f$. Thus, under this assumption $x_l \to x_k|\mathbf{x}$ is equivalent to $\mathbb{V}\{\mu_{k,l}[n]\} > 0$. In particular EIPR vanishes for Granger non-causality ($x_l \nrightarrow x_k|\mathbf{x}$).

Let us finally compare PDC and EIPR in the spectral domain, which underlines the reflections regarding normalization. When expressing EIPR (6.2.5) by the spectral densities of the partial contribution terms and using expression (6.2.7), we obtain

$$\eta^2_{k,l} = \frac{\int_f |A_{k,l}(f)|^2 S_{x_l}(f)\,df}{\int_f |A_{k,k}(f)|^2 S_{x_k}(f)\,df} = \frac{\int_f S_{\mu_{k,l}}(f)\,df}{\int_f S_{\mu_{k,k}}(f)\,df}. \quad (6.2.9)$$

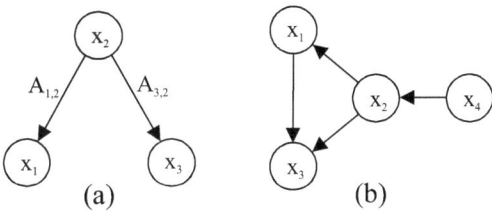

**Figure 6.2.2: Dependence graphs of AR models.** *(a) illustrates a normalization problem of the PDC as discussed in Subsection 6.2.5, $A_{3,2}$ affects $\pi^2_{1,2}$. (b) shows the dependence structure of signal model (6.2.11) used for the assessment of EIPR.*

If we represent PDC (3.4.15) by means of expression (6.2.7), we obtain

$$\pi^2_{k,l}(f) = \frac{|A_{k,l}(f)|^2 S_{x_l}(f)}{\sum_{n=1}^{K} |A_{n,l}(f)|^2 S_{x_l}(f)} = \frac{S_{\mu_{k,l}}(f)}{\sum_{n=1}^{K} S_{\mu_{n,l}}(f)}. \qquad (6.2.10)$$

### 6.2.7  Signal model

As a test case for EIPR and the channel selection algorithm, we consider a simulation based on an example proposed by Winterhalder et al. (2005). This is an autoregressive system of order $p = 5$

$$\begin{cases} x_1[n] = & 0.8\,x_1[n-1] + 0.65\,x_2[n-4] + \varepsilon_1[n] \\ x_2[n] = & 0.6\,x_2[n-1] + 0.6\,x_4[n-5] + \varepsilon_2[n] \\ x_3[n] = & 0.5\,x_3[n-3] - 0.6\,x_1[n-1] + 0.4\,x_2[n-4] + \varepsilon_3[n] \\ x_4[n] = & 1.2\,x_4[n-1] - 0.7\,x_4[n-2] + \varepsilon_4[n] \end{cases} \qquad (6.2.11)$$

with the covariance matrix of the noise set to identity. We simulate 100 seconds assuming a sampling frequency of $f_s = 128$ Hz for consistency with the ECoG data. Note that in this artificial case we process the stationary 5-dimensional signal in one single data window of length $N_{win} = 12800$. The imposed dependency paths of the AR model (6.2.11) are shown in Fig. 6.2.2 (b). This structure was successfully retrieved by application of PDC; compare Winterhalder et al. (2005).

As it is unlikely in applications that one observes values of EIPR exactly matching zero, one has to statistically test whether values of EIPR are significantly different from zero. As no exact distribution of EIPR is available yet, we make use of bootstrapping in order to numerically derive a significance threshold. The idea of the so-called *surrogate data method*[6] is to resample the original data independently for each channel $x_k[n]$ for $N = 100$ times, thus destroying the inter-channel dependence structure. This repetition gives empirical distributions of each EIPR under the null-hypothesis $\mathcal{H}_0$ of non-causality. Here we use the *LOOM (leave-one-out) method* introduced by Schlögl and Supp (2006) for the re-sampling process and subsequent statistical t-test, as it yields reliable results in causal analysis (see Florin et al. (2011) for a comparative study).

In the following let us denote resampled EIPR values by $\tilde{\eta}_{k,l}^2$ ($N$ realizations, index $N$ omitted for reasons of simplicity) and the EIPR value to be tested by $\eta_{k,l}^2$.

Under the assumption that the EIPR values $\tilde{\eta}_{k,l}^2$ (based on the resampled data) are normally distributed under $\mathcal{H}_0$ (with their mean very close to zero), we can employ the well-known t-test by considering the test statistics

$$T = \frac{\sqrt{N}\left(\overline{\tilde{\eta}_{k,l}^2} - \eta_{k,l}^2\right)}{\hat{\sigma}(\tilde{\eta}_{k,l}^2)} \sim t_{N-1;\alpha}. \qquad (6.2.12)$$

Here, the bar $\overline{\phantom{\cdot}}$ denotes the empirical mean and $\hat{\sigma}(\cdot)$ the empirical standard deviation of the EIPR values $\tilde{\eta}_{k,l}^2$ based on the re-sampled data. $t_{N-1;\alpha}$ is the quantile of the Student distribution with $N - 1 = 99$ degrees of freedom and $\alpha = 1 - 0.99 = 0.01$.

Under $\mathcal{H}_0$, $\overline{\tilde{\eta}_{k,l}^2}$ represents the average (resampled) EIPR value in case of non-causality (expected to be very close to zero), and $\eta_{k,l}^2$ needs to be significantly larger to indicate causality. Thus, we consider[7]

$$\begin{cases} \mathcal{H}_0 : \text{non-causality} \\ \mathcal{H}_1 : \text{causality}. \end{cases}$$

---

[6] Compare Maiwald et al. (2008) for a systematic discussion and Kaminski et al. (2001) for an application to DTF.

[7] Note that we employ a one-sided test ($\mathcal{H}_1$: $\eta_{k,l}^2 > \overline{\tilde{\eta}_{k,l}^2}$ rather than $\eta_{k,l}^2 \neq \overline{\tilde{\eta}_{k,l}^2}$) as EIPR values are bigger than zero by definition.

In this test setting we reject $\mathcal{H}_0$ if $\eta_{k,l}^2 > \overline{\tilde{\eta}_{k,l}^2}$, i.e. if $T < t_{N-1;\alpha}$. In other words, the given EIPR $\eta_{k,l}^2$ indicates causality if it exceeds the threshold

$$\eta_{k,l}^2 > -t_{N-1;\alpha} \frac{\hat{\sigma}(\tilde{\eta}_{k,l}^2)}{\sqrt{N}} + \overline{\tilde{\eta}_{k,l}^2}. \qquad (6.2.13)$$

## 6.3 Results

### 6.3.1 Signal model

In order to show the ability of our method to detect dependencies we first apply EIPR to the autoregressive model (6.2.11)[8]. Here we disable the automatic channel selection algorithm described in Subsection 6.2.3 in order to assure that the entire coupling information contained in the multi-channel signal is used.

We compare our findings to the result of PDC as reported by Winterhalder et al. (2005). For each channel $x_l$ and $x_k$ with $k \neq l$, Winterhalder et al. (2005) show a frequency plot of PDC $\pi_{k,l}^2(f)$. These frequency plots are arranged in a $K \times K$-matrix plot, where the columns indicate the source channels and the rows the target channels (compare Fig. 6.4.1 in Subsection 6.4.1). Thus, the (k,l)-subplot quantifies the influence from $x_l$ to $x_k$. If $\pi_{k,l}^2(f) = 0 \,\forall f$, one can conclude that there is no direct dependency from $x_l$ to $x_k$. However, as it is unlikely in applications that one observes values of PDC exactly matching zero for all frequencies, one has to use a statistical test. Thus, Schelter et al. (2005) derived an asymptotic frequency-dependent confidence interval: For each frequency $f$, PDC values below the respective threshold indicate the absence of any direct coupling, compare Subsection 3.4.4.

In contrast to PDC, EIPR condenses the coupling information from $x_l$ to $x_k$ in one scalar value. Therefore, the coupling information (EIPR and significance threshold (6.2.13)) can be represented in a table: In complete

---
[8]Simulation of this model was done with the help of the Matlab® package *arfit* developed by Schneider and Neumaier (2001). In order to avoid initial transient effects, we generate 13800 samples and discard the first 1000.

| $\eta_{k,l}^2$ | $x_1$ | $x_2$ | $x_3$ | $x_4$ |
|---|---|---|---|---|
| $x_1$ | 1.00000 | **0.18039** *(0.05877)* | 0.00005 *(0.00126)* | 0.00001 *(0.00194)* |
| $x_2$ | 0.00066 *(0.00141)* | 1.00000 | 0.00064 *(0.00090)* | **0.75701** *(0.24812)* |
| $x_3$ | **2.09681** *(0.12395)* | **0.26936** *(0.04652)* | 1.00000 | 0.00003 *(0.00133)* |
| $x_4$ | 0.00019 *(0.00031)* | 0.00008 *(0.00017)* | 0.00014 *(0.00025)* | 1.00000 |

**Table 6.3.1:** *Values of EIPR for signal model (6.2.11). Imposed dependencies (bold values) are correctly recognized, 99%-significance thresholds are indicated in italic between brackets.*

analogy to PDC, the (k,l)-element of the table quantifies the influence from $x_l$ to $x_k$, and the columns indicate the source channels, the rows the targets.

As detailed in Table 6.3.1, which is constructed in this way, our measure correctly identifies the imposed dependencies illustrated in Fig. 6.2.2 (b): The ones which are induced by the signal model (6.2.11) are set in bold-face type. Similar to PDC, we do not expect to observe EIPR values exactly matching zero in case of non-causality. We rather have to decide whether EIPR $\eta_{k,l}^2$ significantly differs from zero by exceeding 99% significance thresholds (detailed in italic between brackets behind the respective EIPR values in Table 6.3.1).

### 6.3.2 Analysis of the channel selection algorithm

In a next step we analyze the dynamical channel selection algorithm by applying it to the autoregressive model (6.2.11) without subsequent calculation of EIPR. As stated in Subsection 6.2.3, we expect our algorithm to select the extrinsic channels which contribute significantly to the explanation of the respective intrinsic channel. The simple structure of signal model (6.2.11) allows to verify this design: In this artificial case, the imposed signal model dependencies (see Fig. 6.2.2 (b)) exhaustively define the important extrinsic channels for each intrinsic one. Unlike in the case of ECoG recordings, we do not have any additional weak dependencies here which we want to single out for numerical reasons.

Table 6.3.2 illustrates the results of this simulation. As expected, the algorithm builds up the extrinsic channels sets in accordance with the imposed dependencies:

First, consider channel $x_1$, which is only influenced by $x_2$. The algorithm sets $\mathbb{L}_1 = \{x_2\}$, as in the first step the AIC value of the extended regression using channels $x_1$ and $x_2$ is minimal. In the second step, a further increase of the number of regressors does not lead to a decrease of the information criterion any more, and the algorithm stops returning $\mathbb{L}_1 = \{x_2\}$.

Next, consider channel $x_2$. Similarly to the previous case, it is only influenced by one channel, namely $x_4$, and we obtain $\mathbb{L}_2 = \{x_4\}$.

The situation is different in case of channel $x_3$ which is influenced by $x_1$ as well as by $x_2$. In a first step, the algorithm selects the channel with the strongest influence (AR coefficient of -0.6, see model (6.2.11)), $x_1$. In a second step, $x_2$ is chosen (AR coefficient of 0.4). In a third step, the information criterion cannot be reduced, and the algorithm stops returning $\mathbb{L}_3 = \{x_1, x_2\}$.

Finally, we obtain an empty extrinsic channel set for $x_4$, as the regression based on $x_4$ alone minimizes AIC.

### 6.3.3  Seizure onset zone localization

We apply our proposed methodology, i.e. regression with dynamically selected channels and subsequent calculation of EIPR, to ECoG recordings in order to localize the SOZ. Its identification is based on the analysis of the dependency measure calculated in the initial seconds of the seizure, given the exact seizure onset time (compare Section 4.3).

Data are processed within windows of four seconds, as this value turned out to provide a good trade-off for the estimation quality of the correlation matrix between estimation errors due to instationarity and inaccuracy due to a too small number of samples.

We choose an autoregressive model order of $p = 8$. This allows for the modeling of a spectrum with four peaks, e.g. a prominent peak modeling the rhythmic ictal activity in the $\vartheta$-band and three additional peaks in the other

| $x_k$ | Step | Initial regression: Channels, AIC | Extended regression: Additional channel, AIC | Step result |
|---|---|---|---|---|
| $x_1$ | 1 | $\{x_1\}$: 0.8079 | $x_2$: **-0.0253**<br>$x_3$: 0.7774<br>$x_4$: 0.7491 | choose $x_2$<br>$\mathbb{L}_1 = \{x_2\}$ |
|  | 2 | $\{x_1,x_2\}$: **-0.0253** | $x_3$: -0.0247<br>$x_4$: -0.0246 | STOP<br>$\mathbb{L}_1 = \{x_2\}$ |
| $x_2$ | 1 | $\{x_2\}$: 0.6196 | $x_1$: 0.6182<br>$x_3$: 0.6186<br>$x_4$: **0.0055** | choose $x_4$<br>$\mathbb{L}_2 = \{x_4\}$ |
|  | 2 | $\{x_2,x_4\}$: **0.0055** | $x_1$: 0.0060<br>$x_3$: 0.0060 | STOP<br>$\mathbb{L}_2 = \{x_4\}$ |
| $x_3$ | 1 | $\{x_3\}$: 1.1617 | $x_1$: **0.3708**<br>$x_2$: 0.5860<br>$x_4$: 1.1371 | choose $x_1$<br>$\mathbb{L}_3 = \{x_1\}$ |
|  | 2 | $\{x_3,x_1\}$: 0.3708 | $x_2$: **-0.0075**<br>$x_4$: 0.3395 | choose $x_2$<br>$\mathbb{L}_3 = \{x_1,x_2\}$ |
|  | 3 | $\{x_3,x_1,x_2\}$: **-0.0075** | $x_4$: -0.0068 | STOP<br>$\mathbb{L}_3 = \{x_1,x_2\}$ |
| $x_4$ | 1 | $\{x_4\}$: **0.0113** | $x_1$: 0.0117<br>$x_2$: 0.0119<br>$x_3$: 0.0117 | STOP<br>$\mathbb{L}_4 = \{\}$ |

Table 6.3.2: *Step-wise behavior of the channel selection algorithm for signal model (6.2.11). Channels with imposed dependencies are selected. Minima of the AIC values of each step are set in bold-face type for better traceability.*

physiological frequency bands ($\delta$, $\alpha$, $\beta$). This choice is in good accordance with simulations yielding the optimal model order (Graef 2008).

Furthermore we choose a significance threshold of 0.5, and EIPR values below this cut-off value are discarded. This choice assures that only the strongest couplings are displayed and will be discussed in Subsection 6.4.3.

Fig. 6.3.1 illustrates the results for the three analyzed seizures. Plots (a) - (c) show the MR scan of the patient's head together with the electrode po-

| Seizure | Investigator | Initial Electrodes |
|---|---|---|
| 1 | Algorithm | A7, A10, A11, C2, C5 |
| | Expert 1 | B8 |
| | Expert 2 | A11, A12, B8 |
| | Expert 3 | A10, A11, A12 |
| 2 | Algorithm | A10 |
| | Expert 1 | A11, A12 |
| | Expert 2 | A11, A12 |
| | Expert 3 | A11, A12 |
| 3 | Algorithm | A10 |
| | Expert 1 | A9, A10 |
| | Expert 2 | A9 |
| | Expert 3 | A8, A9 |

**Table 6.3.3:** *SOZ according to causality analysis.* Results in three seizures compared to visual inspection by clinicians.

sitions and results of the EIPR analysis for the respective seizure. An arrow from electrode $x_l$ to electrode $x_k$ indicates that $\eta^2_{k,l}$ exceeds the imposed threshold, i.e. indicates strong directed coupling from $x_l$ to $x_k$. Thus, the arrow maps of Fig. 6.3.1 highlight the areas of increased coupling activity. In all three seizures arrows point away from the parieto-occipital region, in particular from electrode A10. Therefore, this electrode is associated with an area of strong directed dependencies.

In addition to a visual analysis of EIPR arrows (Fig. 6.3.1) one can look at condensed information: The out-degree per channel, defined as the number of arrows pointing away from the channel, is a measure of the directed coupling activity of the respective electrode.[9] Fig. 6.3.2 shows a histogram of out-degrees per channel for each seizure. In seizure 1 (plot (a)), channels A7, A10, A11, C2 and C5 have the highest number of outgoing arrows, in seizures 2 and 3 (plots (b) and (c)) channel A10. Table 6.3.3 summarizes these findings and compares them to the visual analysis of the three clinical experts.

---
[9]We will follow this reasoning in Chapter 7 as well.

These observations suggest that the SOZ comprises the electrode A10, which is in good accordance with the visual analysis of the clinicians. In Subsection 7.4.2 we will discuss this result and related neurophysiological aspects in more detail.

## 6.4 Discussion

### 6.4.1 EIPR as coupling indicator

In this chapter we introduce a novel approach to the quantification of directed couplings. The proposed dependency measure EIPR indicates Granger causality (as does the Partial Directed Coherence), but has the advantage of a clear physical interpretation as a power ratio, compare expression (6.2.5). As mentioned in Subsection 6.2.5, its normalization assures that the measured coupling strength between two channels is not influenced by others. This behavior is in contrast to the one of PDC (Graef et al. 2009) whose confidence level depends on all neighborhood channels in order to compensate for the normalization effect (Schelter et al. 2005).

Due to its construction EIPR successfully validates the signal model (6.2.11). It retrieves the imposed dependencies (compare Table 6.3.1), as does PDC (compare Fig. 6.4.1). In both cases the couplings $x_1 \to x_3$ (position (3,1) in the scalar matrix / matrix plot) and $x_4 \to x_2$ (position (2,4) in the scalar matrix / matrix plot) are predominantly indicated.

Here we want to discuss three additional aspects regarding the comparison of EIPR and PDC:

First, we observe an interesting behavior of EIPR: The statistically significant values in Table 6.3.1 exceed the non-significant ones by a factor of 100. Even at a first glance at such an EIPR table (without comparing the EIPR values to their respective significance thresholds) we would obtain an idea about the underlying dependence structure. Note that the PDC matrix-plot in Fig. 6.4.1 creates a similar impression, but in case of EIPR

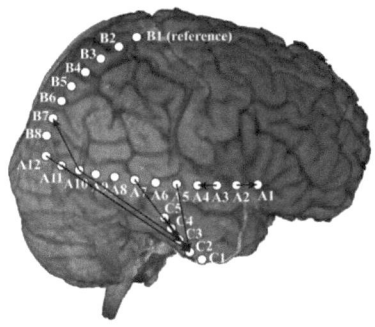

(a) Seizure 1. Data window: 16:12:45 - 16:12:49

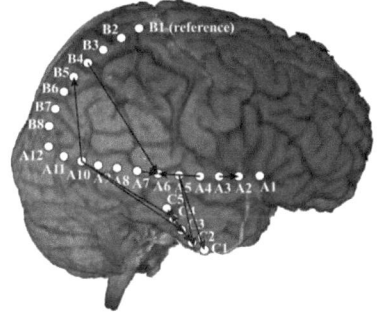

(b) Seizure 2. Data window: 16:48:05 - 16:48:09

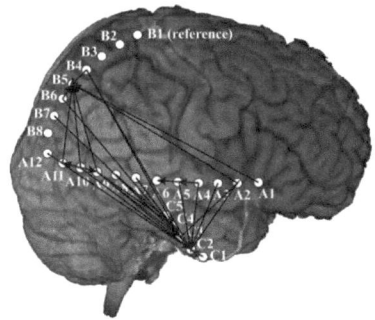

(c) Seizure 3. Data window: 17:18:31 - 17:18:35

**Figure 6.3.1:** *Arrow maps based on EIPR.* Results for the initial four seconds of (a) seizure 1, (b) seizure 2, (c) seizure 3. An arrow between two electrodes indicates an EIPR value above the threshold, i.e. strong directed coupling between respective electrodes.

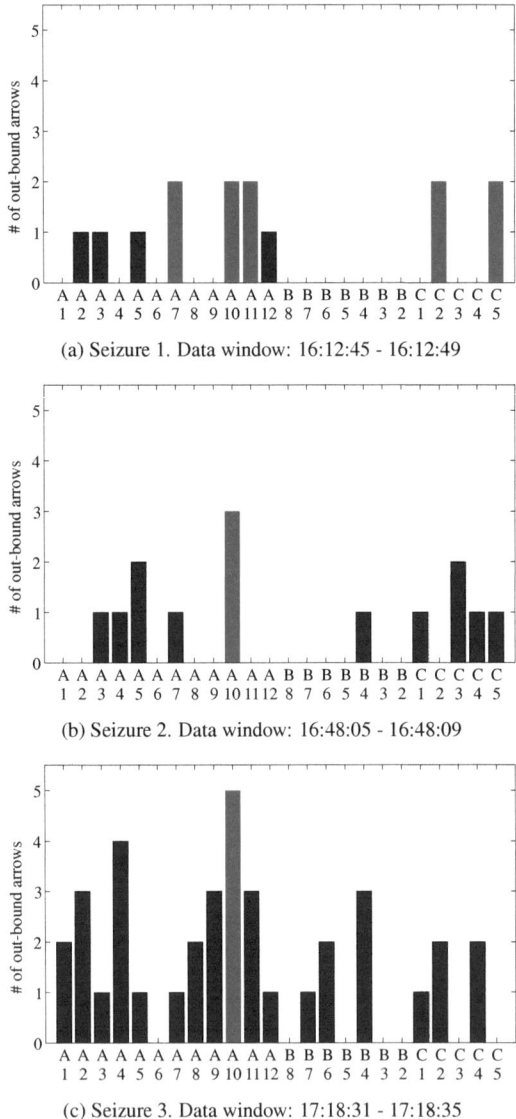

(a) Seizure 1. Data window: 16:12:45 - 16:12:49

(b) Seizure 2. Data window: 16:48:05 - 16:48:09

(c) Seizure 3. Data window: 17:18:31 - 17:18:35

**Figure 6.3.2:** *Out-degrees of arrow maps based on EIPR. Results for the initial four seconds of (a) seizure 1, (b) seizure 2, (c) seizure 3. Highest values of out-degrees marked in red.*

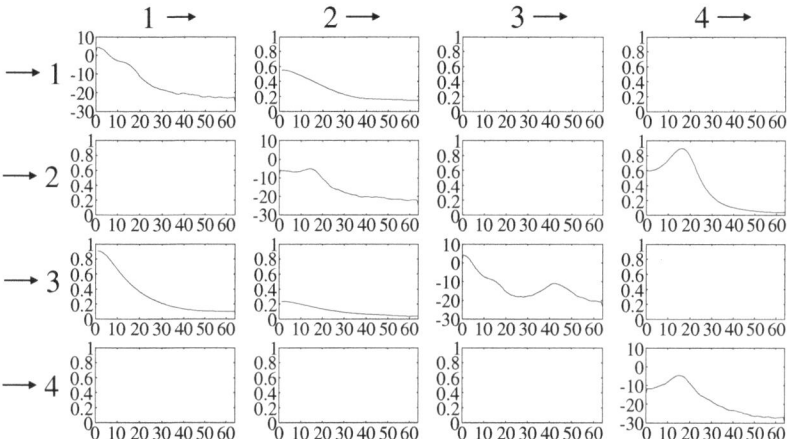

**Figure 6.4.1:** *PDC matrix plot of the signal model (6.2.11)*, confer *(Winterhalder et al. 2005). The plots on the diagonal show the spectra of the respective channels (x-axis: frequency in Hz, y-axis: spectrum in dB scale). A subplot on position $(k,l)$, $k \neq l$ (x-axis: frequency in Hz, y-axis: PDC) visualizes the influence of $x_l$ to $x_k$ measured by $\pi_{k,l}^2(f)$. The imposed dependencies are correctly identified.*

the tendency to separate significant from non-significant values is stronger. This is a result of the normalization discussed above.

Second, the EIPR values in Table 6.3.1 range between 0 and 2, PDC is normalized between 0 and 1. The reason for the scatter of the EIPR values is the following: The variance of the extrinsic contribution term in the numerator of EIPR represents the power of the extrinsic contribution, which is the integral of the corresponding power spectral density over all frequencies (compare expression (6.2.9)). In the numerator of PDC the same integrand shows up, but for a single frequency (compare expression (6.2.10)). Therefore, EIPR takes large values for couplings where PDC is increased over a wide frequency range, compare Fig. 6.4.1. In particular this is the case for the two couplings mentioned above, $x_1 \to x_3$ and $x_4 \to x_2$. Vice versa, PDC vanishing over a large frequency band results in very small EIPR values (e.g. coupling $x_3 \to x_1$).

Third, an advantage of EIPR is its compact representation in form of a matrix of (physically meaningful) scalar values as in Table 6.3.1.

This allows for a simultaneous comparison of the individual EIPR values with their respective significance thresholds even in case of large scale differences. In contrast, a PDC matrix representation has the drawback of being difficult to interpret. One has to consider the respective subplot and compare PDC to the significance threshold for all frequency points. However, as mentioned in (Schelter et al. 2005), this point-wise comparison is not straight-forward. Consider for example the significant couplings $x_2 \to x_3$ and the non-significant ones $x_3 \to x_4$ for both measures. Comparing the EIPR values $\eta_{3,2}^2$ and $\eta_{4,3}^2$ with their respective significance thresholds is easily performed in Table 6.3.1. In case of the PDC matrix plot in Fig. 6.4.1 this simple evaluation is not possible. Small PDC values and significance thresholds are not easily visible due the large scale differences. In order to allow for a clear visualization of the PDC values and their thresholds in each subplot (simultaneously visible), each subplot would have to be scaled differently. Compare Fig. 6.4.2 for an illustration, where the scaling of the two subplots of couplings $x_2 \to x_3$ and $x_3 \to x_4$ is performed in this way. Here, the PDC values and significance thresholds of both couplings are visible, at the price of a scale difference of factor 100. This would render the comparison of the PDC values between different subplots in a matrix plot such as Fig. 6.4.1 difficult.

We want to conclude this part of the discussion with two comments on the interpretation of EIPR.

EIPR is not normalized between 0 and 1, which is a drawback in comparison to PDC. In particular this impairs the comparison between different systems, as equal EIPR values might not indicate the same coupling strength in distinct multi-channel signals.

On the other hand, EIPR allows for an interpretation similar to the signal-to-noise ratio (SNR): Given a signal $x[n] = u[n] + z[n]$ consisting of meaningful information $u[n]$ and background noise $z[n]$, the SNR is commonly defined (Oppenheim and Schafer 1989) on the logarithmic dB

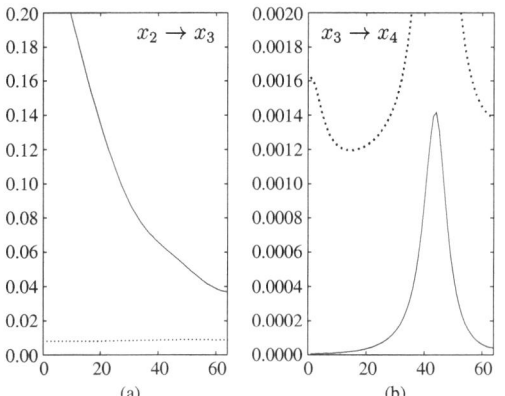

**Figure 6.4.2:** *Zoom into two subplots of the PDC matrix-plot in Fig. 6.4.1 (x-axis: frequency in Hz, y-axis: PDC). (a) zoom of the subplot on position (3,2) indicating the dependence $x_2 \to x_3$, (b) zoom of the subplot on position (4,3) indicating the dependence $x_3 \to x_4$. PDC values are illustrated by solid lines, significance thresholds by dotted lines. For each plot a different zoom factor is necessary to allow for a simultaneous visualization of PDC values and significance thresholds. A direct comparison between the different plots is difficult.*

scale as

$$\text{SNR} \triangleq 10 \lg \left( \frac{\mathbb{V}\{u[n]\}}{\mathbb{V}\{z[n]\}} \right).$$

Thus, EIPR can be intuitively interpreted: The extrinsic contribution takes the roles of the information we are interested in, and the intrinsic contribution is seen as background noise. This interpretation underlines the influence of the extrinsic information for quantifying the coupling strength, which is in particular important in the dependence analysis of epileptic ECoG recordings. We will discuss this setting in Subsection 6.4.3.

## 6.4.2 Behavior of the channel selection algorithm

As demonstrated in Subsection 6.3.2, the dynamic channel selection algorithm behaves as expected in simulations, by selecting channels influencing the others and by discarding other channels. Applied to the signal model (6.2.11), it builds up the respective extrinsic channel sets in accordance with the dependencies imposed. Moreover, it is capable of prioritizing extrinsic channels with strongest influence, compare Table 6.3.2.

This behavior strengthens our conjecture that the algorithm performs well in ECoG data: Here we encounter many influences with few important ones (representing epileptic activity): The order according to which the extrinsic channel set is built up is important, as the proposed forward-selection procedure does not search through the whole parameter space. Due to the simulation results discussed above we are confident that the algorithm's focus on strongest influence selects the important channels first, thus including the channels of interest in the extrinsic channel set. This assures that EIPR can be calculated and visualized between highly coupled channels in the subsequent step.

### 6.4.3 Seizure onset zone localization

In this chapter we assume that the area of highest EIPR values in the initial seconds after seizure onset indicates the SOZ. Our reasoning is the following: As mentioned in Subsection 2.3.5, in case of focal epilepsy the pathological synchronous activity starts at a small localized brain area. Departing from the SOZ it spreads to its immediate vicinity recruiting more and more parts of the neural network. This leads to a hyper-synchronous behavior of the observed channels. One could imagine a »focus« located in the SOZ driving the surrounding channels by imposing its oscillatory frequency in the course of the recruiting process. This could be interpreted as a kind of information transfer: Imagine one electrode in the focus, say $x_1$, influencing the behavior of the surrounding electrodes, say $x_2$ and $x_3$, in the initial phase of the seizure. Then, sticking to this image of information transfer, we expect the extrinsic contributions from $x_1$ to $x_2$ and $x_3$ to show high values and the intrinsic contribution terms of $x_2$ and $x_3$ to be small. This results in high EIPR values $\eta^2_{2,1}$ and $\eta^2_{3,1}$, we observe increased directed coupling activity symbolized by arrows. By limiting our representation to the highest EIPR values within each analysis (significance threshold set to

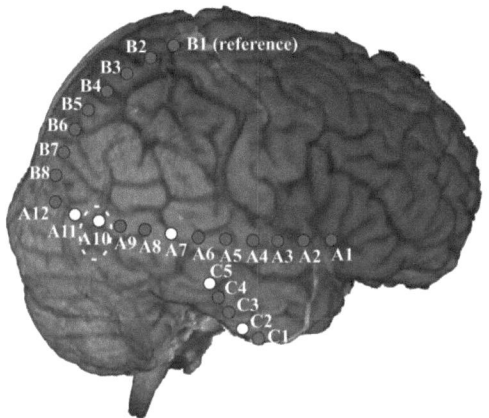

**Figure 6.4.3:** *SOZ according to causality analysis on MRI scan with electrodes positions. Electrodes revealing initial epileptic activity are marked in white (seizure 1: A7, A10, A11, C2, C5; seizures 2 and 3: A10). The supposed seizure onset zone is outlined.*

0.5), we focus on these pathological synchronizations and do not regard others (e.g. weak physiological ones).

Consequently, we expect arrows pointing away from the areas of increased synchronizational activity, in particular from the SOZ, and electrodes of these focus areas to have a high out-degree of EIPR arrows. In order to derive the SOZ, we consider Fig. 6.3.1 for the arrow maps and Fig. 6.3.2 for the corresponding out-degrees of each electrode in the following.

The overall results of our methodology are in good accordance with the SOZ as indicated by the clinicians (Table 6.3.3): In all three seizures electrode A10 is among the channels with strongest coupling activity. While the out-degree histogram is not conclusive in case of seizure 1 (Fig. 6.3.2 (a)), the corresponding arrow map (Fig. 6.3.1 (a)) clearly reveals activity departing from electrode A10 (as well as A11 and A12, both located in the parieto-occipital region). The high out-degrees of electrodes C2 and C5, as shown in Fig. 6.3.2 (a), are a result of EIPR arrows within the C-electrodes ($C2 \to C3, C2 \to C4, C5 \to C2, C5 \to C4$). We assume that strip C captures strong synchronous activity from the hippocampal region located in deeper brain structures below this strip. Obviously this excessive activity leads to feedback mechanisms which results in the observed dependencies on the cortex.[10]

Moreover, synchronous activity departing from A10 can be prominently observed in seizures 2 and 3, see Fig. 6.3.1 (b) - (c). In these cases the out-degree histogram shows conclusive results, compare Fig. 6.3.2 (b) - (c). Taking all these considerations into account, we conclude that the SOZ comprises electrode A10 (marked in Fig. 6.4.3).

### 6.4.4 Concluding remarks

In this chapter we employed a novel dependency measure which is capable of reliably measuring coupling effects in multivariate signals as well as an automatic channel selection algorithm. In particular we are able to identify synchronization effects in ictal multichannel ECoG recordings which allows us to draw conclusions on the localization of the SOZ. We want to conclude this discussion with two side-remarks detailing alternatives:

First, the order of the autoregression is kept constant. As a potential drawback, this might lead to under- or over-fitting of the AR model and consequent erroneous dependencies. In order to avoid this situation one could consider a regression model with order changing over time, see e.g. Prado et al. (2001). In this case the model would better reflect the changing spectral properties of the EEG, but at the price of a higher computational effort: As a function of the data-driven model order a dynamic window length (which is currently fixed) would have to be defined such that a sufficiently reliable estimation of the necessary AR parameters is possible.

Second, we use a dynamic channel selection algorithm (see Fig. 6.2.1) to overcome the estimation issues due to the high correlation in the cross-sectional dimension. Another approach which might be appropriate for this task is penalized regression, e.g. LASSO as introduced by Tibshirani (1996). Only recently, Chiang et al. (2009) successfully applied this approach to

---

[10]In Chapter 7 we observe the same phenomenon by application of influence analysis, see the discussion in Subsection 7.4.2.

neural data, calculated PDC and visualized the indicated brain connectivity of participants taking part in a virtual-reality experiment.

In conclusion, we believe that the aspects discussed in this section strengthen our hypothesis of EIPR being a useful measure for the characterization of neurophysiological dependencies. Therefore, we think that our methodology has the potential to assist clinicians in the presurgical evaluation of epilepsy patients by objectivating the visual ECoG examination: Tracking the synchronization effects over time might indicate the SOZ as well as the initial propagation of the epileptic activity.

# 7 Influence Analysis

> Die Ereignisse der Zukunft können wir nicht aus den gegenwärtigen erschließen. Der Glaube an den Kausalnexus ist der Aberglaube.
>
> — Ludwig Wittgenstein: *Logisch philosophische Abhandlung*
>
> We cannot infer the events of the future from those of the present. Belief in the causal nexus is superstition.
>
> — *Tractatus Logico-Philosophicus*[1] by Ludwig Wittgenstein (1889-1951), Austrian philosopher and pioneer of analytic philosophy

This chapter is based on material which has already been published together with co-workers. We refer to Flamm et al. (2013) for the original article.

## 7.1 Introduction

### 7.1.1 Background

In this chapter we are concerned with the detection of Granger causality in multivariate signals whose components show strong co-movement, i.e. high correlation between the component-series. Following the idea of Chapter 6, we represent these directed dependence relations by arrows in a graph, whose vertices are formed by the channels of the multivariate signal.

As initially proposed by Flamm et al. (2013), we apply a novel methodology for the causal analysis of high-dimensional co-moving data, termed *influence analysis*. It combines factor models (compare Section 3.6) and Granger causality analysis (compare Section 3.3).

---

[1] Sentence 5.1361; English translation by Pears/McGuinness. Original text and English translation accessible online at people.umass.edu/phil335-klement-2/tlp/tlp.html.

While Granger (1969) analyzed the causality between two time series, this concept has been generalized to conditional Granger causality, compare Subsection 3.3.1. Conditional Granger causality, based on autoregressive modeling (compare Section 3.2), is particularly popular in the analysis of neurological signals. For recent applications in neuroscience see e.g. Guo et al. (2008), Liao et al. (2010), Sommerlade et al. (2012) and Flamm et al. (2012).

In practice we often encounter high-dimensional signals with strong co-movement, e.g. in EEG analysis. The naive approach to a Granger causal analysis in this context would be to fit a $K$-dimensional AR model to the $K$-dimensional signal $\mathbf{x}[n]$. We typically encounter two problems:

First, Granger causality analysis is usually considered in the case of regular AR systems, i.e. where the error covariance matrix $\Sigma$ is regular (compare Section 3.2). As EEG data are highly correlated and show strong co-movement, regular AR models lead to a poor estimation and misleading results of the subsequent causality analysis. A visual analysis of ECoG data quickly confirms this co-movement, compare Figs. 4.3.2, 4.3.3 and 4.3.4.

Second, fitting of a $K$-dimensional AR(p) model requires the estimation of $K^2 p$ parameters. In order to obtain reliable estimators for large cross-sectional dimension $K$, a large sample size is required. However, as EEG data show a highly non-stationary behavior (again, compare Figs. 4.3.2, 4.3.3 and 4.3.4), such large sample sizes impair the estimation quality. Again, this leads to poor results of the causality analysis.

### 7.1.2 Granger causality for factor models

In order to avoid these problems we consider factor models, which are a useful tool for EEG analysis (Molenaar 1985, Molenaar and Nesselroade 2001), compare Section 3.6. Naturally the question arises, which causalities can be reasonably analyzed in this context. In this chapter we assume that the dependence of the latent variables $\chi[n]$ properly reflects the causal structure of the observations $\mathbf{x}[n]$. This assumption seems meaningful despite the separation (3.6.1) into noise and latent variables, as will be discussed in Subsection 7.4.1.[2]

The first idea for a causal analysis in the factor model scenario would be to consider relations of the form

$$\chi_i \stackrel{?}{\to} \chi_j | \chi_V, \qquad (7.1.1)$$

where $\chi$ are the latent variables according to Definition 26 in Section 3.6, and the conditioning set is $V = \{1, \ldots, K\}$. However, the usage of the exhaustive set $V$ leads to problems, see Flamm et al. (2013) for technical details. Therefore, we restrict the conditioning set to a sub-set of $V$: Instead of $V$ we use *channel selections* $I \subset V, \#I = q < n$, and we consider relations of the form

$$\chi_i \stackrel{?}{\to} \chi_j | \chi_I \quad i, j \in I. \qquad (7.1.2)$$

The channel selection $I$ has to be chosen appropriately, such that relations of the type (7.1.2) yield reasonable results. This will be discussed briefly in the following[3] (compare Section 3.6 for the notation):

For a channel selection $I \supset i, j$ we consider the corresponding subsystem of (3.6.4)

$$\chi_I[n] = \Lambda_I \mathbf{z}[n] = \Lambda_I \mathbf{A}^{-1}(z) \varepsilon[n], \qquad (7.1.3)$$

where $\Lambda_I$ is the square sub-matrix of $\Lambda$ corresponding to the selected components $\chi_I$.

In order to yield reasonable causal relations of the form (7.1.2), we only consider channel selections $I$ where the corresponding $\Lambda_I$ is regular. We call such $I$ *admissible*.

By rewriting (7.1.3) as an AR representation we obtain

$$\underbrace{\Lambda_I \mathbf{A}(z) \Lambda_I^{-1}}_{\check{\mathbf{A}}(z)} \chi_I[n] = \underbrace{\Lambda_I \varepsilon[n]}_{\check{\varepsilon}[n]} \qquad (7.1.4)$$

---
[2] This assumption is not necessarily satisfied, see Anderson and Deistler (1984).
[3] A detailed derivation can be found in Flamm et al. (2013). For a more theoretical perspective we refer to Flamm (2012).

with $\det(\Lambda_I) \neq 0$. Note that we pre-multiply (7.1.4) with $\Lambda_I$ in order to obtain the leading coefficient of the left-hand side polynomial as the identity, $\check{A}[0] = I_{q \times q}$.

The Granger causality relations of this representation (7.1.4) can now easily be checked according to definition (7) of conditional Granger causality, compare Subsection 3.3.1. In this context, according to (7.1.4), the criterion takes the form

$$\chi_i \not\to \chi_j | \chi_I \iff \check{A}_{ji}(z) = 0 \quad i,j \in I; i \neq j. \tag{7.1.5}$$

Note, however, that the causality relations do depend on the channel selection $I$, compare Flamm et al. (2013).

## 7.2 Method

### 7.2.1 Proposed methodology

Our methodology consists of three steps. First, we use PCA to separate the observations into the latent variables (explaining the co-movement) and the noise. As mentioned in Subsection 7.1.2, we assume that the causal structure of the observations is reflected in the causality structure of the latent variables.

Second, for fixed channel indices $i$ and $j$ we analyze the conditional Granger causality relation $\chi_i \stackrel{?}{\to} \chi_j | \chi_I$, given a fixed channel selection $I \supset i, j$.

Third, we perform this analysis for all admissible channel selections $\tilde{I} \supset i, j$ and derive a heuristic statement for the influence from $\chi_i$ to $\chi_j$, condensing the information of all sub-systems.

In detail we proceed as follows:

First, we perform a PCA on the observations $\mathbf{x}[n]$ in order to obtain the factor loading matrix $\Lambda$ and the static factors $\mathbf{z}[n]$. The dimension of the static factors $q$ is determined via a Scree plot (Cattell 1966). In this graphical method the principal components are sorted according to their explanation of the total variance of the data in descending order. Typically, a bending point can be observed in this graphical representation, which divides the principal components into important and unimportant ones, compare Fig. 7.3.1.

Now let the channel indices $i, j$ and the channel selection $I$ be fixed. The straight-forward application of the approach described in Subsection 7.1.2 yields two problems:

While in theory we can easily distinguish regular and singular matrices $\Lambda_I$ in equation (7.1.4) by considering the determinant, the estimator $\hat{\Lambda}_I$ will typically yield $\det(\hat{\Lambda}_I) \neq 0$. The causality relations drawn from systems with very small values of $|\det(\hat{\Lambda}_I)|$ are not meaningful, which is due to the fact that $\check{\mathbf{A}}(z)$ in (7.1.4) cannot be computed reliably due to the bad conditioning of $\hat{\Lambda}_I$. As the term $|\det(\hat{\Lambda}_I)|$ is a measure for the similarity of the selected channels, we only consider channel selections $I$ with $|\det(\hat{\Lambda}_I)|$ exceeding a threshold $\tau$. This threshold is chosen empirically in order to yield reasonable results.

A similar challenge arises in the estimation of $\hat{\mathbf{A}}_{ji}(z)$. In theory $\check{\mathbf{A}}_{ji}[s] = 0 \ \forall s$ signifies that $\chi_i$ is Granger non-causal for $\chi_j$, recall criterion (7.1.5). However, in estimation we typically have $\hat{\mathbf{A}}_{ji}[s] \neq 0$, so we have to statistically test whether the polynomial coefficients $\hat{\mathbf{A}}_{ji}[s]$ (for all lags $s$) are significantly jointly different from zero.

For this purpose we use an $F$-test ($\mathcal{H}_0 : \check{\mathbf{A}}_{ji}[s] = 0 \ \forall s$), which is implemented in the *GCCA toolbox* and described in Seth (2010). We consider the p-value of the test as a measure for Granger causality: Rejection of $\mathcal{H}_0$ ($p < 0.01$[4]) signifies Granger causality, acceptance means non-causality.

In order to sum up, for each channel selection $I$ (for fixed channel indices $i, j$) we obtain two values: $|\det(\hat{\Lambda}_I)|$ as a similarity measure of the channels in $I$ and the p-value as an indicator for the causality from $\chi_i$ to $\chi_j$.

As a global influence statement from $\chi_i$ to $\chi_j$ is our goal, we want to

---

[4]Note that in Flamm et al. (2013) we use $p < 0.03$. A smaller p-value in this application puts the focus on the highly significant causality relations and leads to fewer directed dependence relations, i.e. clearer arrow maps.

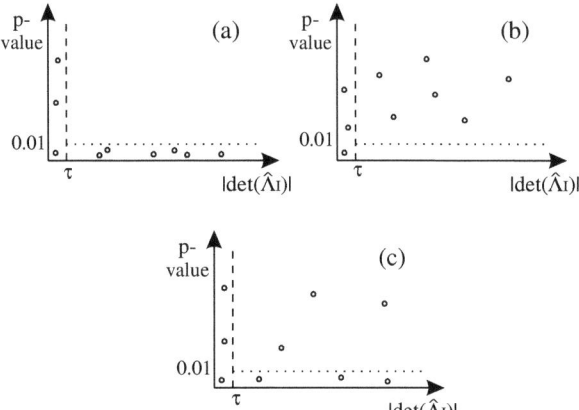

**Figure 7.2.1:** *Visualization of the influence analysis.* Analysis for all causality relations $\chi_i \to \chi_j | \chi_I$ for distinct channel selections $I \supset i, j$. In each plot a point shows the p-value (as a measure of causality) and $|\det(\hat{\Lambda}_I)|$ (as a measure of channel similarity) for the respective channel selection $I$. Points with $|\det(\hat{\Lambda}_I)| > \tau$ only are considered in the analysis (numerical reasons). (a) All relevant points have an associated p-value $< 0.01$, i.e. indicate causality (for each respective $I$). We conclude that $x_i$ influences $x_j$. (b) All relevant points have an associated p-value $> 0.01$, i.e. indicate non-causality (for each respective $I$). We conclude that $x_i$ does not influence $x_j$. (c) For different $I$, causality as well as non-causality statements are indicated. We do not conclude any influence statement.

condense the different conditional causality statements based on distinct channel selections $I$ into a single one. For this purpose we propose an intuitive rule: If all statements for distinct channel selections match, we conclude a global influence statement.

In other words: if $\chi_i \to \chi_j | \chi_I$ for all $I$ with $|\det(\hat{\Lambda}_I)| > \tau$, we say that $\chi_i$ *influences* $\chi_j$. On the other hand, if $\chi_i \not\to \chi_j | \chi_I$ for all $I$ with $|\det(\hat{\Lambda}_I)| > \tau$, we say that $\chi_i$ *does not influence* $\chi_j$. In case of non-conclusive Granger causality statements we do not derive any global influence statement.

Finally, as the causality structures of the observations and the latent variables are assumed to be equal according to Subsection 7.1.2, we say $x_i$ *influences* $x_j$ if $\chi_i$ influences $\chi_j$. The analogous reasoning holds in case of non-influence.

For a better understanding we want to visualize the described methodology: For fixed channel indices $i$, $j$ we plot a point for each distinct channel selection $I \supset i, j$ into the plane spanned by $|\det(\hat{\Lambda}_I)|$ on the x-axis and the p-value on the y-axis. This procedure yields graphs such as shown in Fig. 7.2.1. In such a plot we only consider points with $|\det(\hat{\Lambda}_I)| > \tau$, which are located to the right of the dashed vertical threshold line. Points to the left of this determinant threshold line are ignored, because the corresponding p-values are not meaningful due to numerical instabilities.

A point situated below the dotted line represents a p-value $< 0.01$ and therefore indicates Granger causality. Consequently, a point lying above the dotted line indicates Granger non-causality.

Fig. 7.2.1, where each plot is constructed as described above, illustrates the three cases we distinguish:

In plot (a) all relevant points are situated below the dotted line, i.e. each point individually indicates causality ($\mathcal{H}_0$ of non-causality rejected due to $p < 0.01$), thus we have global influence.

We observe the opposite situation in plot (b), where all relevant points are above the dotted line, i.e. each point individually indicates non-causality, so we speak of global non-influence.

Plot (c) illustrates a situation where distinct channel selections lead to Granger causality as well as Granger non-causality statements. In this case, we refrain from concluding on global influence.

## 7.2.2 Signal model

In order to assess the proposed methodology we apply it to simulated data where we know the imposed dependence structure. Consider the following signal model

$$\mathbf{x}[n] = \Lambda \mathbf{z}[n] + \eta[n] \qquad (7.2.1)$$
$$\mathbf{A}(z)\mathbf{z}[n] = \varepsilon[n].$$

First we simulate the 3-dimensional static factors $\mathbf{z}[n]$ as an AR(2) process with

$$\mathbf{A}(z) = \begin{pmatrix} 1 - 0.2z & 0 & 0 \\ -0.3z^2 & 1 - 0.5z & 0 \\ -0.7z^2 & 0 & 1 - 0.5z \end{pmatrix}$$

and the covariance matrix of the noise set to identity.[5]

For the construction of $\mathbf{x}[n]$ we choose the factor loading matrix

$$\Lambda = \begin{pmatrix} 1 & 0 & 0 \\ 0 & 1 & 0 \\ 0 & 0 & 1 \\ 1 & 0 & 0 \\ 0 & 1 & 0 \\ 0 & 0 & 1 \\ 1 & 0 & 0 \\ 0 & 1 & 0 \\ 0 & 0 & 1 \end{pmatrix}$$

and the variance of the noise

$$\text{Cov}(\eta[n]) = \text{diag}(0.15, 0.15, 0.61, 1.37, 0.61, 0.15, 1.37, 1.37, 0.61).$$

The Granger causality structure of the simulated 3-dimensional static factors $(z_{1*}, z_{2*}, z_{3*})^T$ is depicted in Fig. 7.2.2 (a), the resulting influence structure of the 9-dimensional co-moving system $(x_1, \ldots, x_9)^T$ in Fig. 7.2.2 (b). Note that the simple design of $\Lambda$ yields the clear graph in plot (b).

## 7.3 Results

### 7.3.1 Signal model

We apply our methodology to the simulated data from the signal model (7.2.1).

---

[5] Simulation is done using the function *arsim* of the Matlab package *arfit*, described in Schneider and Neumaier (2001).

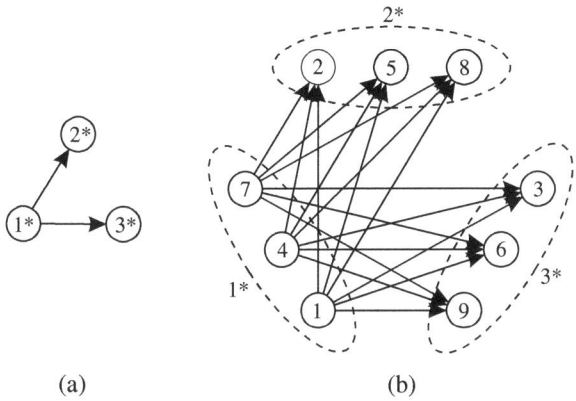

**Figure 7.2.2:** *Illustration of the dependence structure of signal model (7.2.1). (a) Granger causality structure of the simulated static factors (arrows indicate conditional Granger causality). (b) Influence structure of the simulated observations (arrows indicate influence).*

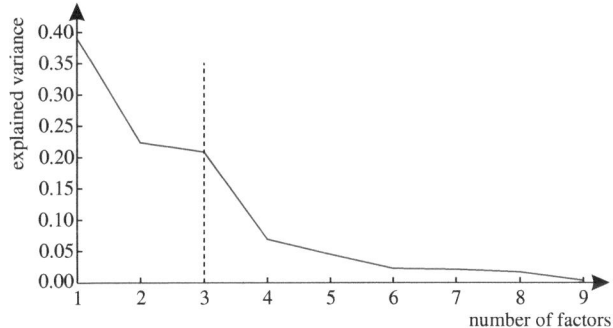

**Figure 7.3.1:** *Scree plot of the Principal Component Analysis of the simulated data from signal model (7.2.1). Three factors explain the majority of the variance.*

For the initial calculation of the PCA, we determine the number of static factors $q$ by considering the Scree plot, see Fig. 7.3.1. This figure shows the percentage of the explained variance per factor. We observe that three factors explain the majority of the variance, thus we choose $q = 3$. Furthermore, by application of BIC (Schwarz 1978) we obtain an AR-model order of $p = 2$ (matching the imposed model order).

Proceeding according to our methodology, for fixed source channel

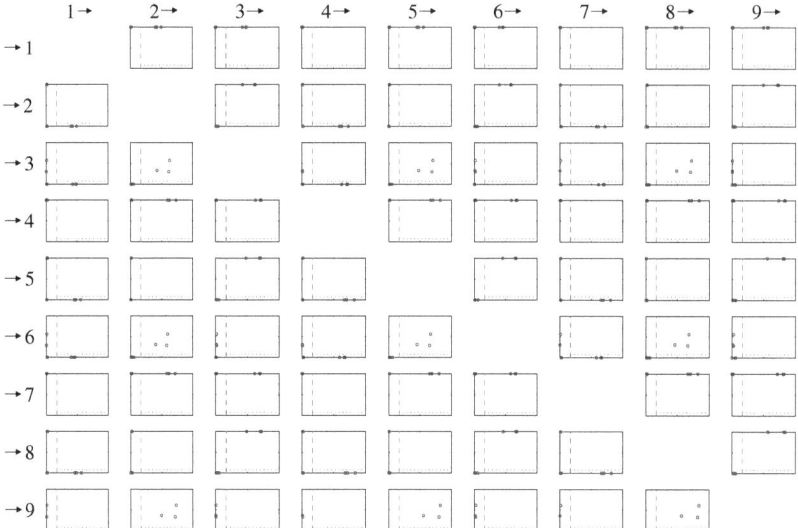

**Figure 7.3.2:** *Results of the influence analysis of signal model (7.2.1) in matrix plot form.* Columns indicate the source channels $x_i$ and the rows the target channels $x_j$, the $(j,i)$-subplot quantifies the influence from $x_i$ to $x_j$. Only points to the right of the dashed threshold line are considered due to numerical reasons (admissible points). If all admissible points are located under the dotted line we say $x_i$ influences $x_j$, compare the interpretation in Fig. 7.2.1. This analysis yields the influence structure illustrated in Fig. 7.2.2 (b).

index $i$ and target channel index $j$ we obtain causality relations for all channel selections $I \supset i, j$. They are represented as points in a graph as described in Subsection 7.1.2. Compare Fig. 7.2.1 for the idea. Hereby points with p-values $> 0.4$ are displayed with p-value $= 0.4$ (i.e. maximum of ordinate), because this does not change the results of the analysis (highly non-significant anyway) and facilitates the visualization.

In Fig. 7.3.2 all these plots are arranged in a $9 \times 9$ matrix plot, where the columns indicate the source channels $x_i$ and the rows the target channels $x_j$. Clearly, the $(j,i)$-subplot quantifies the influence from $x_i$ to $x_j$. Obviously, diagonal elements are not displayed.

Let us consider the interpretation of three selected subplots in Fig. 7.3.2 in detail:

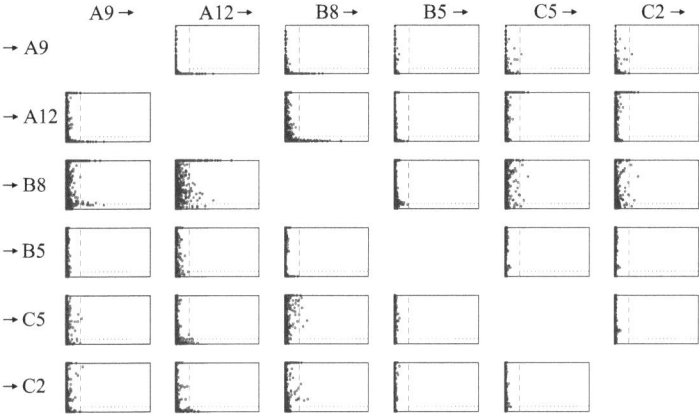

**Figure 7.3.3:** *Results of the influence analysis of seizure 1. Six selected channels in matrix plot form (submatrix of the full $24 \times 24$ matrix). Compare Fig. 7.3.2.*

In subplot (3,1) all points to the right of the determinant threshold line are located below the dotted line, and therefore represent p-values smaller than 0.01 (i.e. the null hypothesis of non-causality is rejected). This means that for all admissible channel selections $I$, we have $\chi_1 \to \chi_3 | \chi_I$. Thus, we say $x_1$ influences $x_3$.

In subplot (3,2) all points to the right of the determinant threshold line are located above the dotted line. Thus, we say $x_2$ does not influence $x_3$.

In subplot (4,1) all points are located to the left of the determinant threshold line, therefore we do not draw any conclusions. The reason for this behavior is that $x_1$ and $x_4$ are both generated by $z_{1*}$ and therefore are highly correlated.

Thus, we have shown that we successfully retrieve the imposed dependence structure of (7.2.1) by interpreting each of the subplots in the described way. Channel $x_1$ influences channels $x_2, x_3, x_5, x_6, x_8, x_9$, so do channels $x_4$ and $x_7$. This is illustrated in Fig. 7.2.2 (b) where each influence relation is symbolized by an arrow.

### 7.3.2 Seizure onset zone localization

We apply the proposed method to the three seizures described in Section 4.3. We process a 4-second segment from the initial phase of each seizure in order to locate the SOZ, compare Figs. 4.3.2, 4.3.3 and 4.3.4.

The number of static factors is determined by a Scree plot as before in Subsection 7.3.1. In order to achieve an explained variance greater than 80%, we choose $q = 5$.

In accordance with the survey paper Tseng et al. (1995), we choose the AR model order for the Granger causal analysis $p = 8$. This allows for the modeling of four spectral peaks, e.g. a prominent peak modeling the rhythmic ictal activity in the $\vartheta$-band and three additional peaks in the other physiological frequency bands ($\delta$, $\alpha$, $\beta$).[6]

Furthermore, we choose $\tau = 0.05$ in order to discard points of the causality analysis which yield unreasonable results due to numerical problems caused by channel similarity.

In analogy to the analysis of the signal model (7.2.1) in Subsection 7.3.1 we obtain a $24 \times 24$ matrix plot for each of the three seizures. For better visibility, Fig. 7.3.3 shows only a $6 \times 6$ sub-matrix of the matrix plot of seizure 1. Here we briefly want to discuss 4 subplots of Fig. 7.3.3, i.e. of seizure 1, in detail:

Subplot (2,3) describes the causality relations from B8 to A12. All points to the right of the determinant threshold are located below the dotted line, thus we say channel B8 influences channel A12.

In subplot (5,3) all points located to the right of the determinant threshold are above the dotted line. Therefore, B8 does not influence C5.

An interesting case occurs in subplot (3,2). We have admissible points above and below the dotted line. In this case we refrain from any influence statement.

Finally in subplot (4,2) all points are located to the left of the determinant threshold. We do not draw any conclusion in this case, as there are no

---
[6]Compare Chapter 6, where we identically choose the AR model order following this line of thought.

| Seizure | Investigator | Initial Electrodes |
|---|---|---|
| 1 | Algorithm | A10, A12 |
|   | Expert 1 | B8 |
|   | Expert 2 | A11, A12, B8 |
|   | Expert 4 | A10, A11, A12 |
| 2 | Algorithm | A5, A6 |
|   | Expert 1 | A11, A12 |
|   | Expert 2 | A11, A12 |
|   | Expert 4 | A11, A12 |
| 3 | Algorithm | A6 |
|   | Expert 1 | A9, A10 |
|   | Expert 2 | A9 |
|   | Expert 4 | A8, A9 |

**Table 7.3.1:** *SOZ according to influence analysis. Results in three seizures compared to visual inspection by clinicians.*

admissible channel selections.

By interpreting each subplot in the $24 \times 24$ matrix of seizure 1 in this way, we obtain all influence relations. In particular we are interested in the (source, target) channel pairs, where the source does influence the target channel.

Fig. 7.3.4 illustrates the results for the three analyzed seizures. Plots (a) - (c) show the MR scan of the patient's head together with the electrode positions and results of the influence analysis for the respective seizure. An arrow from electrode $x_l$ to electrode $x_k$ indicates influence. Thus, the arrow maps of Fig. 6.3.1 highlight the areas of increased coupling activity: in all three seizures the parieto-occiptial region and in seizures 2 and 3 (plots (b) and (c)) the temporal region covered by electrode strip C.

In analogy to Chapter 6, we additionally consider the out-degree per channel as a condensed measure of influence. Fig. 7.3.5 shows a histogram of out-degrees per channel for each seizure. In seizure 1 (plot (a)), channels A10 and A12 have the highest number of outgoing arrows, in seizure 2 (plot (b)) channels A5 and A6 and in seizure 3 (plot (c)) channel A6. Table 7.3.1

summarizes these findings and compares them to the visual analysis of the three clinical experts.

These observations suggest that the SOZ comprises the temporo-parieto-occipital region between electrodes A12 and A5, which is in good accordance with the visual analysis of the clinicians. In Subsection 7.4.2 we will discuss this result and related neurophysiological aspects in more detail.

## 7.4 Discussion

### 7.4.1 Influence analysis

A key assumption of the employed method is that the latent variables reflect the causality structure of the observations. In other words, that we can infer the dependencies between the channels from the latent variables, and that the noise is assumed not to contain any causality information. Although this is a strong assumption, it seems reasonable to us: This assumption is very similar to the one that the causal dependencies are reflected by distinct, high-amplitude wave forms in the signal, e.g. high amplitudes in ictal ECoG recordings. We believe that in particular in neurophysiological applications this is meaningful, as we expect these high-amplitude oscillations to carry substantial information about the causality structure of the generating cortical mechanisms.

A naturally arising question is whether and to which extent the definition of influence is meaningful. In our opinion it is a workable procedure for causality analysis in high-dimensional co-moving systems: Intuitively one expects a certain kind of dependence if $\chi_i$ is causal for $\chi_j$ for all (admissible) channel selections. The opposite case, i.e. no dependence despite causality statements for all (admissible) channel selections, would be somehow surprising.

A potential weakness of the influence definition is the fact that in practical applications one is often confronted with the case where no influence statement can be inferred. Compare subplot (3,2) in Fig. 7.3.3, where

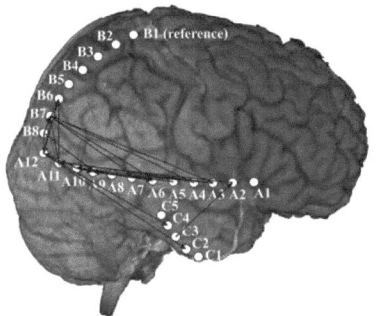

(a) Seizure 1. Data window: 16:12:45 - 16:12:49

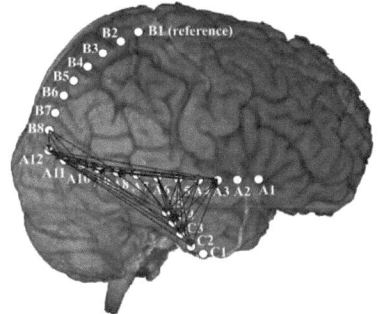

(b) Seizure 2. Data window: 16:48:05 - 16:48:09

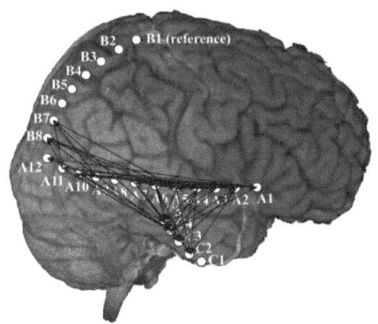

(c) Seizure 3. Data window: 17:18:31 - 17:18:35

**Figure 7.3.4:** *Arrow maps based on influence analysis. Results for the initial four seconds of (a) seizure 1, (b) seizure 2, (c) seizure 3. An arrow between two electrodes indicates an influence statement.*

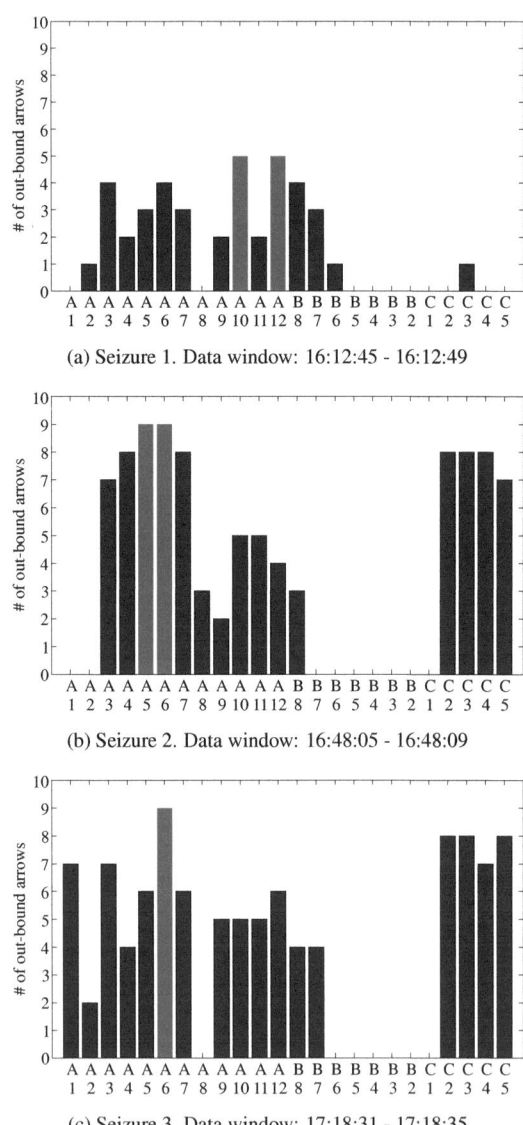

**Figure 7.3.5:** *Out-degrees of arrow maps based on EIPR.* Results for the initial four seconds of (a) seizure 1, (b) seizure 2, (c) seizure 3. Highest values of out-degrees marked in red.

causality as well as non-causality relations are symbolized. In such cases we recommend a more precise investigation which particular channel selections yield causality and which do not (Flamm et al. 2013). Due to the existence of such undecidable cases we avoid the term *causality* and refer to the derived dependence statement as *influence*.

Finally we briefly want to discuss some technical aspects of the employed method, compare Flamm et al. (2013) for further details. As mentioned in Subsection 7.2.1, influence analysis methodology consists of three steps: PCA, Granger causality analysis for fixed channel selection $I$ and derivation of an influence statement. This modular design allows for an easy adaptation of each single step, i.e. alternative methods could be used in each step independently of the others.

First, the usage of sparse PCA would enforce additional zeros in the factor loading matrix $\Lambda$. Thurstone (1947) suggested five criteria for a simple structure and d'Aspremont et al. (2007) gave a direct formulation for sparse PCA.

Second, two central indicators are used in the Granger causality analysis for a fixed channel selection: $|\det(\hat{\Lambda}_I)|$ as a measure of channel similarity (as mentioned above) and the $p$-value of an $F$-test as an indicator for Granger causality. The latter is employed due to its well-established theory. Note that in neuroscience literature various other directed dependency measures are used, compare Section 3.4.

Third, influence statements are derived according to an intuitive rule: If $\chi_i \to \chi_j | \chi_I$ for all admissible $I$, we say that $x_i$ influences $x_j$. However, one could imagine other rules depending on specific applications. For instance, only the channel selection with the largest $|\det(\hat{\Lambda}_I)|$ could be taken into account for the influence statement. Another possible rule for influence statements might be based on the comparison of the number of points above and below the dotted line, e.g. the majority or a certain percentage. This would reduce the cases where no influence statement can be derived based on the current rule, compare e.g. subplot (3,2) in Fig. 7.3.3.

### 7.4.2 Seizure onset zone localization

In this chapter we assume that the area of increased coupling activity in the initial seconds after seizure onset, indicated by influence statements, indicates the SOZ. Our reasoning is the following:

In case of focal epilepsy, the pathological synchronous activity (characterizing the epileptic seizure) starts at a circumscribed brain area. From there, ictal activity spreads to its immediate vicinity recruiting more and more parts of the neural network, compare Subsection 2.3.5. This leads to distinct co-movement of the observations. One could imagine a »focus« located in the SOZ driving the surrounding channels by imposing its oscillatory frequency in the course of the recruiting process. This could be interpreted as a kind of information transfer or causal interaction: The electrodes in the focus influence the behavior of the surrounding electrodes in the initial phase of the seizure. Therefore, we expect to obtain indications for the SOZ by applying a Granger-causal analysis to factor models within the initial seconds of the seizure.

In the first seizure (Fig. 7.3.5 (a)), the electrodes identified by our algorithm (A10, A12) are comprised in the set of initial electrodes specified by the clinical experts (Table 8.3.1). In this case, our findings correlate very well with the visual inspection which strengthens the argumentation above. Interestingly, in seizures 2 and 3 the algorithmic analysis points out electrodes A5 and A6 (Figs. 7.3.5 (a) and (b), respectively), thus suggests a more anterior region as SOZ. This localization differs from the visual analysis (Table 8.3.1), indicating electrodes A11 and A12 in seizure 2 and electrodes A8 to A10 in seizure 3. Note, however, that this trend to more anterior parts is confirmed by the visual analysis. In seizure 3 the SOZ identified by clinical experts is localized in a more anterior region than in seizure 1 (A8-A10), though at a slightly more posterior position (ca. 3cm) than according to our analysis.

In the course of the recruiting process we obviously expect feedback mechanisms between the channels (besides unidirectional dependence).

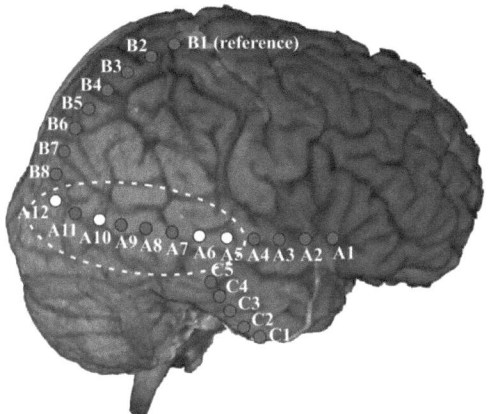

**Figure 7.4.1:** *SOZ according to influence analysis on MRI scan with electrodes positions. Electrodes revealing initial epileptic activity are marked in white (seizure 1: A10, A12; seizure 2: A5, A6; seizure 3: A6). The supposed SOZ is outlined.*

This can be observed in Fig. 7.3.4 (a), consider e.g. channels $B6 \leftrightarrow A12$. However, in the SOZ the departing arrows predominate, i.e. we have channels with a high out-degree.

This situation is reflected in the out-degree histogram in Fig. 7.3.5. First, consider plot (a) for seizure 1. Channels with the highest out-degrees coincide with the SOZ, and with increasing distance to the SOZ the respective out-degree decreases. Channel A8 with an out-degree of zero is an exception, as we cannot infer any influence statement for this source channel (only non-admissible channel selections for all target channels).

Next, consider plots (b) and (c) for seizures 2 and 3. Here, we observe the same situation of decreasing out-degrees as before in seizure 1 (additionally, we cannot infer any influence statements for channel A8 again). The electrodes of strip C represent an exception, as we have large influence from the electrodes of strip C. This might be due to the fact that this electrode strip captures highly synchronized activity from the hippocampal region.[7]

Taking all these observations into account, we conclude that the SOZ

---

[7] Compare Chapter 6, where we also observe this behavior, although applying a different causality analysis method (see the discussion in Subsection 6.4.3).

comprises the temporo-parieto-occipital region between electrodes A12 and A5 (marked in Fig. 7.4.1).

### 7.4.3 Concluding remarks

In this chapter we employed a causal analysis of high-dimensional co-moving data termed influence analysis, connecting the topics of Granger causality and factor models. Influence analysis allows for an investigation of the dependence structure of highly correlated data in neuroscience such as EEG. We would like to conclude this discussion with three remarks:

First, the determination of $q$, i.e. the dimension of the static factors, is very important for the simulated data as well as the ECoG recordings and potentially has great influence on the analysis. In both applications we chose $q$ based on Scree plots in a reasonable way. A systematic way of choosing $q$, depending on the data (e.g. in preictal periods, at the beginning of the seizure and in later phases), has to be established in future work.

Second, a major problem of the presented causal analysis of the ECoG data is stationarity. As can be seen in Figs. 4.3.2, 4.3.3 and 4.3.4, the analyzed ECoG data do not seem to be stationary. During the main phase of the epileptic seizure the channels show distinct rhythmic behavior, which is stationary in nature. However, we are interested in the beginning of the epileptic seizure, where the rhythmic activity starts to spread. Due to this evolution our period of interest is non-stationary. There are two main approaches to cope with the problem of causal analysis of non-stationary data: The first is to use non-linear methods (Marinazzo et al. 2011), the second is to employ the shortest possible time window for the analysis. We focus on the second approach and use a short time window.

Finally, choosing the correct AR model order is not straight-forward. The problem is that already filtering in the preprocessing step of the data could affect the AR-model order and subsequently the Granger-causal analysis of the data (Barnett and Seth 2011). However, we chose an AR-model of constant order $p = 8$, which is in good accordance with simulation yielding the optimal model order of ictal ECoG data (Graef 2008).

Concluding, our methodology correctly identifies the dependence structure of the signal model (7.2.1) and yields promising results for the analyzed ECoG data. Therefore, we think that our methodology might have the potential to assist clinicians in the presurgical evaluation by objectivating their visual ECoG examination.

# 8 Segmentation

> Omnis ars naturae imitatio est; itaque quod de universo dicebam ad haec transfer quae ab homine facienda sunt.
>
> — Lucius Annaeus Seneca: *Epistulae morales ad Lucilium*
>
> All art is but imitation of nature; therefore, let me apply these statements of general principles to the things which have to be made by man.
>
> — Moral letters to Lucilius[1] by Seneca the Younger (ca. 4 BC - 65), Roman philosopher and statesman

This chapter is based on material which has already been published together with co-workers. We refer to Graef et al. (2012) for the original article.

## 8.1 Introduction

### 8.1.1 Background

In this chapter we are concerned with the analysis of early seizure spread based on the spatio-temporal tracking of ictal rhythmic $\vartheta$-activity. For this purpose we employ a methodology which consists of a segmentation step, as proposed by Graef et al. (2012), and of a subsequent, clinically inspired classification step.

Like many other bio-signals, EEG recordings are highly instationary. As this violates the commonly used assumption of stationarity, one has to model the non-stationary character explicitly in order to allow for appropriate signal processing. Segmentation of the signal into stationary fragments is a

---

[1]Letter LXV: On the first cause, paragraph 3; English translation by Richard Mott Gummere. Latin text accesible online at www.intratext.com/IXT/LAT0230.

possible solution: After definition of the segment boundaries, estimation can be performed within each segment by means of well-known techniques.

The straight-forward implementation is a sliding window of a fixed length, wherein the signal is assumed to be stationary. Due to the inherent simplicity of this implementation, in particular regarding the coding effort, this approach is very popular in EEG processing.[2] However, one has to find a trade-off between window length and estimation precision in this case, as the continuously moving window ignores the signal properties: The longer the window, the higher the risk of temporally smeared statistics; the shorter the window, the poorer the estimation quality due to a too low number of signal samples. Therefore, one is interested in a more sophisticated segmentation which should ideally take account of the underlying signal characteristics.

A prominent approach is to consider the temporal evolution of appropriate statistics. Whenever such a statistic exceeds an imposed threshold, one assumes a significant change in the signal characteristics and defines a boundary. The definition of such statistics heavily depends on the application. As Table 8.1.1 details, various statistics have been proposed for segmentation.[3] A prominent method is based on the spectral properties of the signal, the so-called *spectral error measure segmentation* (*SEM*, Bodenstein and Praetorius (1977), Praetorius et al. (1977)) with modifications Bodenstein et al. (1985) and a bivariate extension Gath et al. (1992).

Segmentation based on the temporal evolution of statistics can be accomplished according to three different strategies (Keogh et al. 2004): top-down, bottom-up and sliding window (compare Table 8.1.1). The top-down strategy starts with the entire data window, breaks it up into non-overlapping segments and repeats this procedure for each segment iteratively, until the segmentation is so fine that no significant statistical changes can be found any more. The bottom-up strategy operates in the inverse way. It initially

---

[2] For instance, this approach is pursued in Chapters 6 and 7.
[3] Note that recently the idea of the so-called *2nd-stage-tracking* (Varsavsky and Mareels 2006, 2007) came up: not the statistic itself reflects the underlying signal appropriately, but its changes. Following this line of though, the variation of the differentiated statistics has to be considered.

| Type | Statistics (»feature«) | Top-down approach | Bottom-up approach | Sliding-window approach |
|---|---|---|---|---|
| Direct physical interpretability | Spectral content | | Adak (1998), Carré and Fernandez-Maloigne (1998) | Bodenstein and Praetorius (1977), Praetorius et al. (1977), Bodenstein et al. (1985), Gath et al. (1992), Lavielle (1993) |
| | Mean | Fukuda et al. (2004) | | |
| | Teager energy according to Kaiser (1990) | | | Wu and Gotman (1998), Agarwal and Gotman (1999) |
| | Line length | | | Esteller et al. (2001) |
| No direct physical interpretability | Kolmogorov-Smirnov statistics | Brodsky et al. (1999) | | Kaplan et al. (2001), Kaplan et al. (2005) |
| | Itakura distance | | | Kong et al. (1995), Kong et al. (1997) |
| | AIC minimization | | Inouye et al. (1995) | |
| Classical non-linear methodology, difficult to interpret | Statistical dimension | | Celka and Colditz (2002) | |
| | Non-linear correlation coefficient | | Terrien et al. (2008) | |
| | Phase synchronization | | Terrien et al. (2008) | |

**Table 8.1.1:** *Overview of segmentation strategies. Common statistics.*

defines a fine grid of potential segment boundaries and iteratively joins parts with the same statistical properties until joining is no longer possible due to a lack of significant statistical changes. The sliding window strategy compares the statistical properties of two adjacent windows and defines a boundary in case of a significant difference between them. This comparison can happen in different forms (Chu 1995): a fixed reference window and a sliding test window of the same size (used in this chapter, compare Subsection 8.2.1), a growing reference window and an adjacent test window of fixed size, a global fixed reference window and a growing test window, or a growing reference windows and an adjacent shrinking test window.

Another approach for segmenting is the minimization of a cost function, whereby the segmentation and the estimation happen in parallel in an iterative way. An exemplary list of algorithms of this class includes Lavielle (1998), the SLEX algorithm in its bivariate (Ombao et al. 2001) and multivariate form (Ombao et al. 2005) and the Auto-PARM-algorithm (Davis et al. 2006).

We finally mention alternative EEG segmentation approaches including such different methods as hidden Markov models (Cassidy and Brown (2002), Penny and Roberts (1999)) or adaptive estimation with parameter changes defining the boundaries, e.g the »forgetting« factor in Ursulean and Lazar (2007), the errors in Lopatka et al. (2005), or the coefficients themselves in Ohlsson et al. (2010).

## 8.1.2 Contribution

In this chapter we apply a physiologically motivated computational approach, which is based on the propagation of rhythmic $\vartheta$-activity: We use a segmentation method based on the relative frequency contributions of ictal ECoG data, as outlined by Graef et al. (2012) (see Subsection 8.2.1). Based on this segmentation we classify each segment with respect to its epileptic character (see Subsection 8.2.3). The temporal delay of the start of epileptic activity on different channels is an indicator for seizure propagation, thus revealing the SOZ (see Subsection 8.2.4).

## 8.2 Method

Our proposed methodology consists of two consecutive steps, the segmentation of the ECoG data (see Subsection 8.2.1) and the subsequent classification of the resulting segments regarding their epileptic character (see Subsection 8.2.3). Segmentation and classification are applied channel-wise. We will demonstrate segmentation and classification for a single channel $x_k[n]$ in the following.

For the segmentation step (see Subsection 8.2.1) we will calculate a sequence of power spectral densities, for the classification step (see Subsection 8.2.3) we use a rhythmicity analysis.

### 8.2.1 Segmentation

We initially compute sequences of power spectral densities varying over time by using a sliding window of length $T_{\text{win}}$, where the window is moved by $T_{\text{res}}$ seconds. For each window the corresponding spectrum is indexed with the center point of the window. Thus, we obtain a sequence of power spectral densities $S(f)[m]$ with window index $m$ (new temporal resolution $1/T_{\text{res}}$ Hz). Power spectral densities are estimated using the non-parametric Welch method (128-point FFT), compare Subsection 3.1.4.

We use the following five neurophysiologically meaningful frequency bands: $\delta_{low}$ (1.0-1.5 Hz), $\delta_{up}$ (2.0-3.5 Hz), $\vartheta$ (4.0-8.5 Hz), $\alpha$ (9.0-13.5 Hz), $\beta$ (14.0-30.0 Hz). For each time step $m$, we calculate the power within these bands, e.g. $P_\alpha[m] = \int_{9.0}^{13.5} S(f)[m]\,df$, as well as the total power of all these bands, denoted by $P[m]$. The split of the $\delta$-band into a lower and upper part was done in order to separate low-frequency artifacts from physiological activity.

Our segmentation method is based on the temporal evolution of statistics by using a sliding-window approach (compare Subsection 8.1.1). As initially proposed in Graef et al. (2012), we consider the temporal changes of the relative power contribution of the individual physiological frequency bands. These changes are reflected by a a novel statistic termed *Band Power Measure* (BPM$[m]$) defined as

$$\text{BPM}[m] = \left(\frac{P_{\delta low}[m]}{P[m]} - \frac{P_{\delta low}[m^*]}{P[m^*]}\right)^2 + \ldots \quad (8.2.1)$$
$$\ldots + \left(\frac{P_{\beta}[m]}{P[m]} - \frac{P_{\beta}[m^*]}{P[m^*]}\right)^2.$$

The sliding-window algorithm employed is as follows: We choose an initial reference point $m^*$ (preictally), and for increasing $m > m^*$ we calculate BPM$[m]$. If the Band Power Measure exceeds a given threshold $th$, i.e. BPM$[m] > th$, we start a new segment by updating the reference point $m^* = m+1$ and continue the calculation for increasing $m$ until the end of the dataset. The set of reference points obtained by this algorithm yields the boundary points for our segments, i.e. each resulting segment is limited by two subsequent reference points.

The definition of BPM is physiologically motivated: According to Foldvary et al. (2001) ictal EEG (as well as ECoG) in temporal lobe epilepsy patients is often characterized by its distinct rhythmic $\vartheta$- or $\delta$-activity. Due to its definition, BPM ignores small contribution shifts within a frequency band, but is sensitive to frequency shifts from one frequency band into another.[4] We expect a new segment at the beginning of distinct $\vartheta$-activity, i.e. in particular at the beginning of ictal periods.

### 8.2.2 Periodic waveform analysis

In the second step we decide whether a segment represents epileptic activity, as will be detailed in Subsection 8.2.3.

---

[4]Note that this behavior is in contrast to the SEM-based segmentation in Bodenstein and Praetorius (1977), which regards power shifts at arbitrary frequencies, independently of the physiological bands.

This quantification is heavily based on a methodology of rhythmicity measurement termed periodic waveform analysis (PWA), compare Hartmann et al. (2011). In brief, the PWA consists of three steps:

First, the total harmonic energy at time point $n$

$$E_\tau[n] \triangleq \sum_{k>0} \left| \frac{1}{\sqrt{\tau}} \sum_{n'=-\infty}^{\infty} x[n']\, \psi[n'-n] \exp(-2i\pi \frac{kn'}{\tau}) \right|^2 \qquad (8.2.2)$$

is calculated for all cycle durations $\tau$ with $\tau_{min} \leq \tau \leq \tau_{max}$, where $\psi$ is a window of bounded energy centered around 0 of length $\alpha/\tau$.

For fixed $n$, a maximization of (8.2.2) yields the dominant cycle duration $\hat{\tau} = \arg\max_\tau E_\tau$. For our calculations we will use the *coupled frequency* $f_{\text{PWI}}$, which is the inverse of the dominant cycle duration.

Second, the signal energy corresponding to a cycle duration $\tau$ is given by

$$N_\tau[n] \triangleq \frac{1}{\sqrt{\tau}} \sum_{n'=-\infty}^{\infty} \left(x[n']\, \psi[n'-n]\right)^2. \qquad (8.2.3)$$

Finally, for dominant cycle duration $\hat{\tau}$ (or coupled frequency $f_{\text{PWI}}$), the *Periodic Waveform Index (PWI)* is defined as

$$\text{PWI}[n] \triangleq \frac{E_{\hat{\tau}}[n]}{N_{\hat{\tau}}[n]}, \qquad (8.2.4)$$

which quantifies the rhythmic character of a signal. By construction, PWI equals one for perfectly rhythmic signals and is near zero for totally arhythmic ones like white noise.

Thus, we obtain a pair (PWI, $f_{\text{PWI}}$) for each time step, which indicates the rhythmic character of the signal as well as the dominant frequency.

Note that PWI calculations are performed independently of the estimation of the power spectral sequence. For technical reasons (the iterative maximization of (8.2.2) is time consuming) the output of the PWI framework is limited to half the temporal resolution of the power spectral sequence in this study.

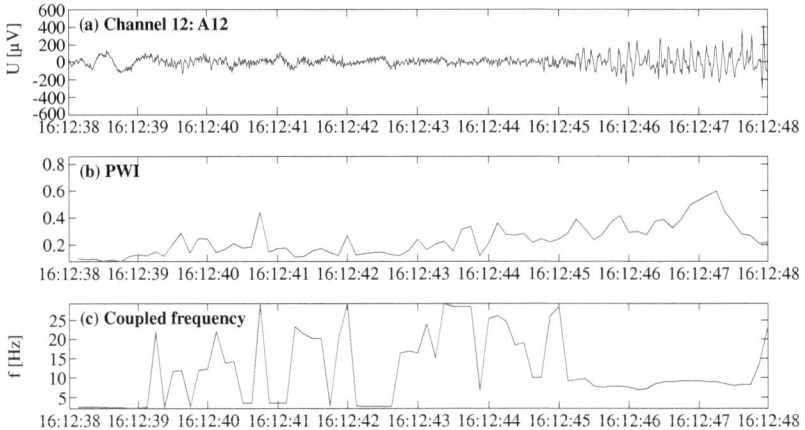

**Figure 8.2.1:** *PWI and coupled frequency* in initial 10 seconds of seizure 1, channel A12. (a) ECoG data, (b) PWI, (c) $f_{PWI}$. Note the unstable dominant frequency in case of low PWI and its convergence to the predominant frequency with increasing rhythmicity.

In Fig. 8.2.1 we see the characteristic behavior of PWI and its coupled frequency, which will be exploited in Subsection 8.2.3: In ictal phases the signal becomes rhythmic, which is reflected by a high PWI value and a stable coupled frequency. Compare Fig. 8.2.1 from 16:12:45 onward, when PWI increases and the coupled frequency converges to the predominant frequency in the $\vartheta$-band. In interictal phases we observe a low PWI, and the coupled frequency oscillates in an unstable manner, which is a result of the numerical maximization of (8.2.2).

### 8.2.3 Segment classification

The core aspect of the methodology in this chapter is the characterization of epileptic activity, which is clinically motivated. The classification of segments regarding their epileptic character is based on the behavior of PWI and its coupled frequency, as shown in Fig. 8.2.1. It follows a two-step approach by imitating the way how epileptologists (at Neurological Center Rosenhügel) perform the visual analysis of early seizure spread:

1. Initial classification.

    In the initial step of their visual analysis clinicians search for periods with distinct epileptic activity. Thus, we also first concentrate on segments showing clear epileptic activity. We classify segments as epileptic (»step-1-positive«) which exhibit a rhythmic character, a stable coupled frequency in the $\delta/\vartheta/\alpha$-bands (or super-harmonics with a base frequency in this range) and high amplitudes. Furthermore, in order to avoid false-positives, we suppress a classification as step-1-positive if rhythmic low-frequency artifacts are present (maximal contribution in the $\delta_{low}$-band).

2. Subsequent classification of segments preceding the epileptiform ones.

    In the second step clinicians search for initial epileptiform discharges in the ECoG. They start from the period of clear epileptic activity (step 1) and scan the preceding electrophysiological activity until they discover the first discharge with a morphological pattern similar to the subsequent clear epileptic activity. Again, our algorithm imitates the clinical procedure and classifies segments preceding step1-positive segments. These preceding segments typically exhibit a rhythmic character and a similar morphology (to the subsequent step1-positive segment, i.e the same spectral properties), but low amplitudes and an unstable coupled frequency. This classification (»step-2-positive«) may be applied to segments immediately preceding step1-positive segments (the segment itself must not be step1-positive).

3. Subsequent classification of segments succeeding the epileptiform ones.

    Finally, note that in the course of a seizure the predominant frequency might change, or activity in other frequency bands (e.g. typically $\beta$-activity in tonic-clonic seizures) could superpose the initial one. Segments showing this behavior have not been marked as epileptic so far, as they typically show high rhythmicity and large amplitudes, but not necessarily a stable coupled frequency. Therefore, we classify segments with high PWI and large amplitudes as epileptic in this step. This classification (»step-3-positive«) applies to each segment immediately succeeding a step1-positive segment (the segment itself must neither be step1- nor step-2-positive).

The rule sets of these three steps (4 rules for step 1, 2 rules for step 2, 2 rules for step 3) as well as their implementation are summarized in Table 8.2.1. As detailed, thresholds are calculated from a 20 min-lasting preictal reference period prior to the first seizure.[5] Thresholds without any direct neurophysiological interpretation have been determined empirically. If all criteria of the respective step are satisfied (step 1: 4; step 2: 2; step 3: 2), we classify a segment as epileptic.

Note that robust statistics are used in order to compensate for the instationary character of the bio-signals: median instead of mean, mean absolute deviation (MAD) instead of standard deviation, inter-quartile range (IQR).

### 8.2.4 SOZ localization and seizure propagation

In order to draw conclusions on the initial seizure spread we apply the segmentation of Subsection 8.2.1 and the subsequent classification of Subsection 8.2.3 channel-wise. The temporal delay of the start of epileptic activity on different channels reflects seizure propagation. The first channels showing epileptic activity indicate the SOZ.

### 8.2.5 Signal model

In order to to illustrate our method of tracking initial seizure propagation, we consider the segmentation of simulated data in Subsection 8.3.1. For this purpose we construct a 10s-lasting test signal $(x_1, x_2, x_3)^T$ with a sampling

---

[5]Unlike in Chapter 5, the reference period for the analysis of all three seizures lies before onset of the first seizure, not before onset of the respective seizure. This assures that the reference values obtained from the 20 min period are not biased by postictal phenomena (e.g. curve suppression), as the three seizures follow each other in a 30 min interval.

| Step | Rule | Criterion | Implementation |
|---|---|---|---|
| Initial classification of segments with distinct epileptic activity | #1 | Rhythmicity: high PWI | median(PWI in segment) > median(ref. PWI) + MAD(ref. PWI) |
| | #2 | Stable coupled frequency | IQR(freq. in segment) < 7 Hz |
| | #3 | Dominant frequency physiologically meaningful (i.e. $\delta/\vartheta/\alpha$) OR dominant frequency is first harmonic of a fundamental frequency in $\delta/\vartheta/\alpha$-range | median($f_{PWI}$ in segment) < 13 Hz OR {median($f_{PWI}$ in segment) < 26 Hz AND 1st PSD peak = median($f_{PWI}$ in segment) +/- 1Hz AND 2nd PSD peak = median($f_{PWI}$ in segment)/2 +/- 0.5 Hz} |
| | #4 | High amplitudes | median(signal envelope in segment) > 90% percentile(ref. signal envelope) |
| Classification of precedent segments | #5 | Rhythmicity: high PWI | median(PWI in segment) > median(ref. PWI) + MAD(ref. PWI) |
| | #6 | Spectral properties similar to the subsequent segment marked in step 1 (i.e. spectral peak in same band OR max. rel. power contribution in same band) | mode(freq. band with spectral peak in segment) = mode(freq. band with spectral peak in subsequent step 1-positive segment) OR mode(freq. band with max. rel. contr. in segment) = mode(freq. band with max. rel. contr. in subsequent step 1-positive segment) |
| Classification of subsequent segments | #7 | Rhythmicity: high PWI | median(PWI in segment) > median(ref. PWI) + MAD(ref. PWI) |
| | #8 | High amplitudes | median(signal envelope in segment) > 90% percentile(ref. signal envelope) |

**Table 8.2.1:** *Segment classification.* *3 subsequent steps with respective rule sets detailing criteria and implementation. Thresholds are calculated from a 20 min-lasting preictal reference period.*

frequency of 128 Hz (for better comparison with ECoG data),

$$\begin{aligned} x_1[n] &= a_1 \sin\left(\frac{2\pi n}{128} f_1\right) + a_2 \sin\left(\frac{2\pi n}{128} f_2\right) + \varepsilon_1[n] \\ x_2[n] &= a_3 \sin\left(\frac{2\pi n}{128} f_3\right) + a_4 \sin\left(\frac{2\pi n}{128} f_4\right) + \varepsilon_2[n], \quad (8.2.5) \\ x_3[n] &= a_5 \sin\left(\frac{2\pi n}{128} f_5\right) + a_6 \sin\left(\frac{2\pi n}{128} f_6\right) + \varepsilon_3[n] \end{aligned}$$

where $(\varepsilon_1, \varepsilon_2, \varepsilon_3) \sim \mathcal{N}(0, 0.04)$ is Gaussian white noise. In order to simulate an epileptic seizure and its propagation, the amplitudes $a_1, \ldots, a_6$ and frequencies $f_1, \ldots, f_6$ follow the temporal scheme

$x_1$ : phase A (0-5 s):   $a_1 = 50$, $f_1 = 15\,\text{Hz}$; $a_2 = 10$, $f_2 = 3\,\text{Hz}$
      phase B (5-10 s):   $a_1 = 50$, $f_1 = 8\,\text{Hz}$; $a_2 = 20$, $f_2 = 20\,\text{Hz}$
$x_2$ : phase A (0-2 s):   $a_3 = 50$, $f_3 = 15\,\text{Hz}$; $a_4 = 10$, $f_4 = 3\,\text{Hz}$
      phase C (2-10 s):   $a_3 = 80$, $f_3 = 10\,\text{Hz}$; $a_4 = 30$, $f_4 = 20\,\text{Hz}$
$x_3$ : phase A (0-3 s):   $a_5 = 50$, $f_5 = 15\,\text{Hz}$; $a_6 = 10$, $f_6 = 3\,\text{Hz}$
      phase B (3-7 s):   $a_5 = 50$, $f_5 = 8\,\text{Hz}$; $a_6 = 20$, $f_6 = 20\,\text{Hz}$
      phase C (7-10 s):   $a_5 = 80$, $f_5 = 10\,\text{Hz}$; $a_6 = 30$, $f_6 = 20\,\text{Hz}$

Hereby, phase A represents preictal activity in the $\beta$-band and phases B and C ictal activity in the $\vartheta$- and $\alpha$-bands, respectively. For more realistic simulation results, the dominant activity ($f_1$, $f_3$, $f_5$) in preictal phase A is superposed with low-amplitude $\delta$-activity ($f_2$, e.g. artifacts) and in ictal phases B and C with low-amplitude $\beta$-activity ($f_4$, $f_6$, e.g. muscle activity). Compare Fig. 8.3.1 (a) for an illustration.

## 8.3 Results

As we are interested in the initial spread of the rhythmic activity, we investigate the first 20 seconds of each of the three seizures, compare Figs. 4.3.2 - 4.3.4. We start at the time indicated by the clinicians, i.e at onset of paroxysmal fast activity (30 Hz) or high-frequency oscillations (78 Hz) in

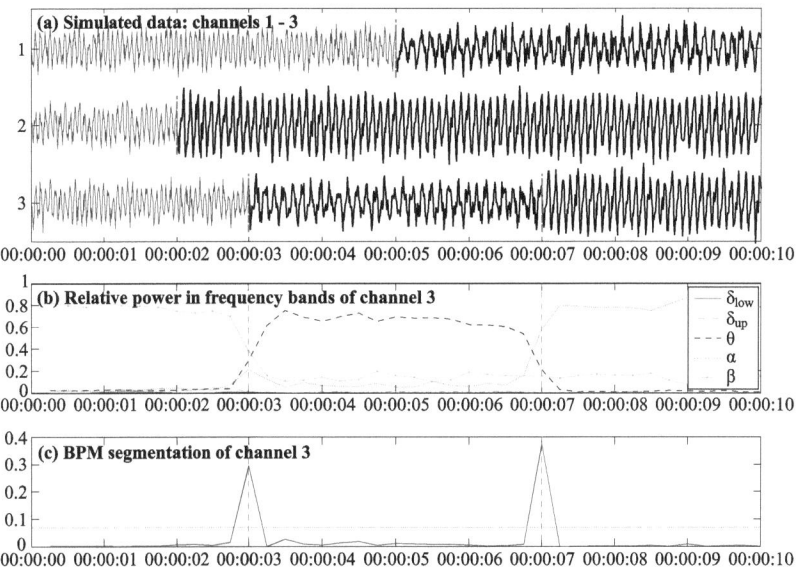

**Figure 8.3.1:** *Segmentation of simulated data (8.2.5). (a) Simulated channels 1- 3, phase A (preictal) marked in blue, phases B and C (both ictal) in black. (b) Relative contributions of the physiological frequency bands of channel 3, (c) Band Power Measure and imposed threshold (dotted) of channel 3. Segment boundaries are indicated by vertical lines in all three graphs. Segmentation corresponds to the imposed temporal structure of the simulated signal and illustrates the tracking of initial seizure propagation (channel 2: onset, channels 3 and 1: follow-up).*

the particular case of seizures 1 and 2. Rhythmic $\vartheta$-activity starts approximately 10 seconds later. Note that the detection of the initial fast activity is not in the focus of this study.[6]

### 8.3.1 Segmentation of simulated data

First we apply the BPM segmentation to the simulated test signal (8.2.5).

Fig. 8.3.1 displays the three simulated channels on top (plot (a)). In each channel, phase A (simulation of preictal activity) is marked in blue and phases B and C (simulation of ictal activity) in black. Details for channel 3 are given in the subplots below: the temporal evolution of the relative

---

[6]See Chapter 5 for HFO detection.

frequency contributions in plot (b) and the corresponding BPM statistics (8.2.1) in plot (c). Segment boundaries are marked in all three graphs to simplify their comparison. The imposed threshold is indicated by a dotted line in plot (c). For consistency with the segmentation results of ECoG data, we use the threshold th $= 0.07$.

The segmentation obtained by the BPM algorithm matches the structural change of the signal very well. The BPM statistic reacts immediately to changes in the frequency content, and segment boundaries are set at the time point following a threshold exceeding. The resulting segmentation of the three-dimensional test signal exactly reflects the imposed temporal structure of the three phases, compare Fig. 8.3.1 (a).

These results illustrate the idea of tracking initial seizure propagation, as outlined in Subsection 8.2.4. We observe the first segment representing ictal activity on channel 2, followed by channels 3 and 1. In this artificial case we conclude that channel 2 indicates the onset of epileptic activity, and initial seizure spread propagates to channel 3 a second later, followed by channel 1 with a delay of two seconds.

### 8.3.2 Segmentation of ECoG data

We apply the proposed methodology with the following set of parameters: $T_{win} = 1.5\,\text{s}$, $T_{res} = \frac{1}{16}\,\text{s}$ for high temporal resolution and reactive segmentation. The initial reference point $m^*$ is set at $0.75$ s to avoid initial transient effects due to preprocessing steps. Furthermore we employ an empirically determined threshold th $= 0.07$, which turned out to represent a good trade-off between segment length and segmentation reactivity for ECoG data, compare Graef et al. (2012).

Fig. 8.3.2 displays the segmentation of an exemplary channel, A12, of seizure 1 in detail. In analogy to the segmentation of the test signal, the ECoG data are shown on top in plot (a). For a better comprehension the temporal evolution of the relative frequency contributions is detailed in the middle (plot (b)) and the corresponding BPM statistics at the bottom (plot

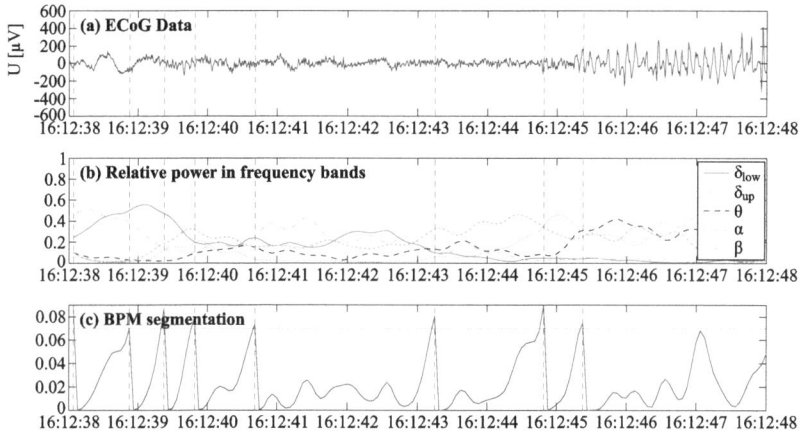

**Figure 8.3.2:** *Segmentation of channel A12, seizure 1. (a) ECoG data, (b) Relative contributions of the physiological frequency bands, (c) Band Power Measure and imposed threshold (dotted). Segment boundaries are indicated by vertical lines in all three graphs.*

(c)). Again, segment boundaries are marked in all three graphs to simplify their comparison, and the imposed threshold is indicated by a dotted line.

In this example a significant change of the BPM statistics can be observed because of frequency shifts from one physiological band into another. Furthermore the segments coincide well with phases of $\vartheta$-activity (16:12:45 - 16:12:48).

### 8.3.3 Classification of ECoG segments

We exemplarily show the mechanism of segment classification for channel A12 of seizure 1 (first 10 seconds). For this purpose Fig. 8.3.3 displays the step 1 segment classification based on rules #1 - #4, compare Table 8.2.1. This figure shows the temporal evolution of the statistics for these four criteria: Again, the ECoG data are shown in plot (a). Bold lines in plots (b) - (e) signify a fulfilled criterion. For a segment to be step 1-positive, criteria #1 - #4 have to be fulfilled.

In this example only the last segment is classified as epileptic (corresponding ECoG data highlighted in black). Note that this segment represents

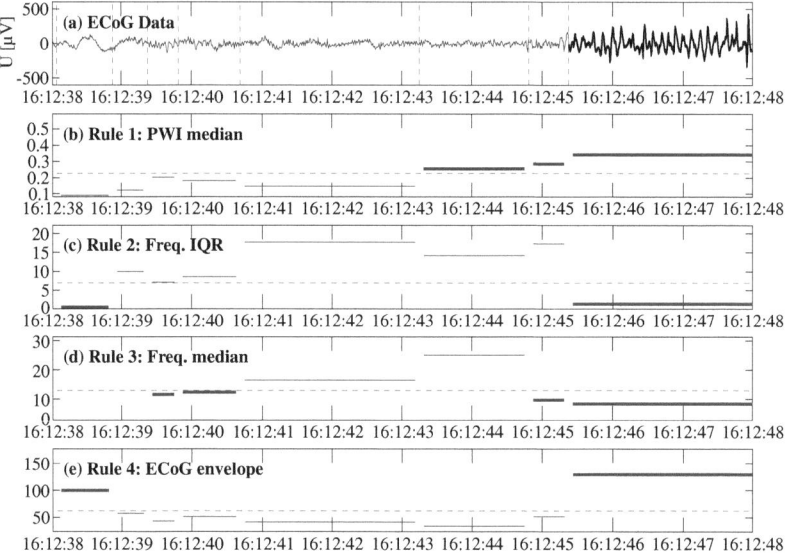

**Figure 8.3.3:** *Segment classification of channel A12, seizure 1 according to step 1. (a) ECoG data with indicated segment boundaries, (b) rule #1: median of PWI in segment, (c) rule #2: IQR of $f_{PWI}$ in segment, (d) rule #3: median of $f_{PWI}$ in segment, (e) median of signal envelope in segment. Thresholds are indicated by horizontal dashed lines. Bold lines of statistics #1 - #4 signify fulfilled respective criterion within segment. A segment is classified as step 1-positive, if all 4 rules are fulfilled, only the last segment is classified.*

the start of the rhythmic epileptic activity on this channel. In preceding segments at least one of the four criteria correctly circumvents a step 1-positive classification.

### 8.3.4 Onset zone analysis

We apply the proposed segmentation and classification procedure to the first 20 seconds of each of the three seizures. For a better visualization of the results we only present a five-second zoom in Fig. 8.3.4. Step 1-positive segments are marked in black, step 2-positive segments in yellow and step 3-positive segments in magenta. Onset times according to the three clinical experts, who individually analyzed the raw ECoG data, are indicated above the respective channels.

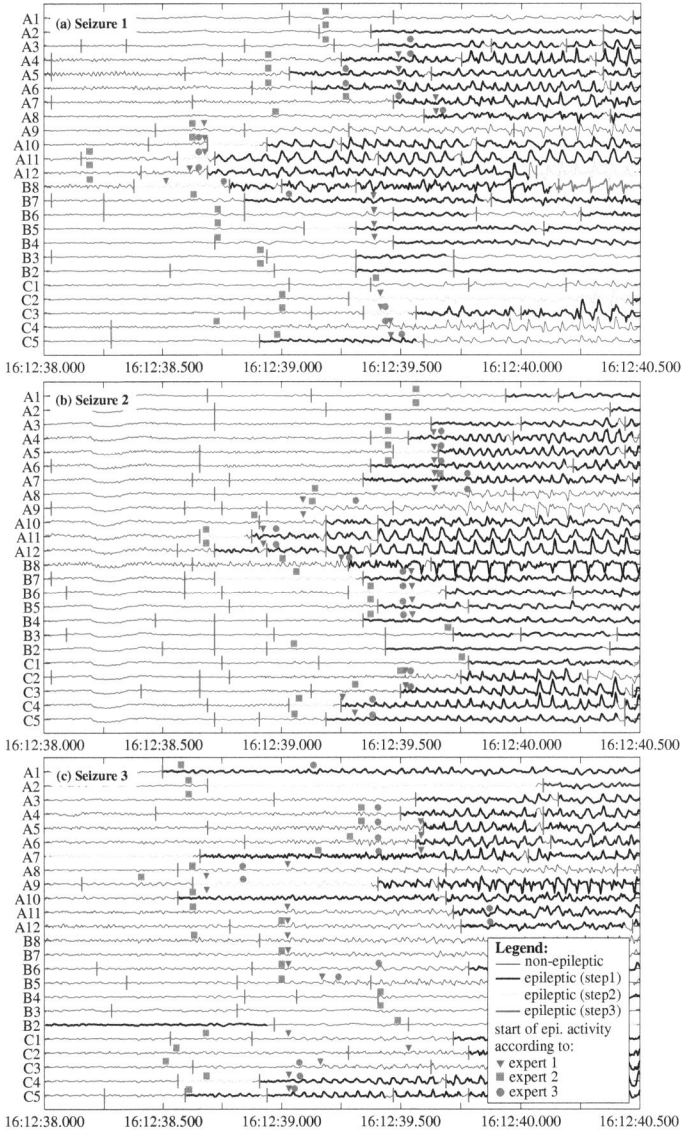

**Figure 8.3.4:** *SOZ and initial spread of epileptic activity.* *(a) Seizure 1, (b) seizure 2, (c) seizure 3. Onset according to clinical experts marked on respective channels. Segment boundaries are indicated by vertical red lines. Segments with distinct epileptic activity (step 1-classification) are marked in black, step 2-classification in yellow, step 3-classification in magenta.*

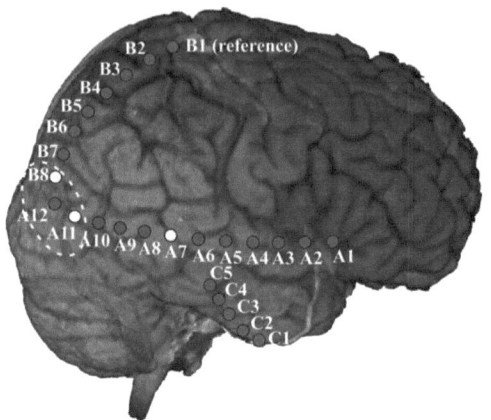

**Figure 8.3.5:** *SOZ according to segmentation method on MRI scan with electrodes positions. Electrodes revealing initial ictal HFO activity are marked in white (seizure 1: B7, B8; seizure 2: A9). The supposed SOZ is outlined.*

Table 8.3.1 summarizes our findings as well as the visual analysis. According to our algorithm, the rhythmic epileptic activity starts on channel B8 (follow-up: A10, A11, A12) in seizure 1, on A11 (follow-up: A12, B7) in seizure 2 and on A7 (follow-up: A1, A9, A10, C4, C5) in seizure 3. As can be seen from Fig. 8.3.4, our results correlate well with the visual analysis of the clinical experts.

Based on these findings (analysis of initial electrodes) we infer that the SOZ comprises the temporo-parieto-occipital area between electrodes B8 and A7. We will discuss a refinement of the SOZ to the area of electrodes A11 to B8 in Subsection 8.4.3, compare the MRI scan in Fig. 8.3.5 for a visualization of the electrode positions.

## 8.4 Discussion

### 8.4.1 Segmentation

In Subsection 8.3.2 we showed that the BPM-based segmentation is an appropriate method for the segmentation of ictal ECoG data. Here we

| Seizure | Investigator | Initial Electrodes | Close follow-up |
|---|---|---|---|
| 1 | *Algorithm* | *B8* | *A10, A11, A12* |
|  | Expert 1 | B8 | A10, A11, A12 |
|  | Expert 2 | A11, A12, B8 | A9, A10, B7 |
|  | Expert 3 | A10, A11, A12 | B8 |
| 2 | *Algorithm* | *A11* | *A12, B7* |
|  | Expert 1 | A11, A12 | A9, A10 |
|  | Expert 2 | A11, A12 | A10 |
|  | Expert 3 | A11, A12 | B8 |
| 3 | *Algorithm* | *A7* | *A1, A9, A10, C4, C5* |
|  | Expert 1 | A9, A10 | A8, A11, A12, B6, B7, B8, C1, C4, C5 |
|  | Expert 2 | A9 | A1, A2, A3, C2, C3 |
|  | Expert 3 | A8, A9 | A1, C3, C4, C5 |

**Table 8.3.1:** *SOZ and initial propagation according to segmentation method. Results in three seizures compared to visual inspection by clinicians.*

want to discuss two additional aspects for seizure 1, the behavior of the segmentation algorithm in interictal and ictal periods and the influence of the threshold on the segment length.

First, consider the behavior of the segmentation algorithm in interictal and ictal periods. Fig. 8.3.2 underlines the advantages of the construction of the BPM statistics: Prior to the rhythmic $\vartheta$-activity (starting at ca. 16:12:45) we observe quickly interchanging frequency contributions, see plot (b). This results in a BPM statistics with high variations and frequent threshold exceeding, see plot (c). Therefore, our algorithm yields short segments in this period. On the other hand, during the rhythmic activity, only small power shifts occur within the physiological frequency bands. Thus, the frequency contribution of the respective bands show a constant behavior, namely the $\vartheta$-band on a high level, which results in longer segments.

Fig. 8.4.1 shows a comparison of segment lengths preictally and ictally for seizure 1. For this analysis, we consider one minute prior to the onset of rhythmic $\vartheta$-activity (at 16:12:45) and the initial minute of this rhythmic activity. As Fig. 8.4.1 reveals, segment lengths are longer in the ictal period (plot (b)) than in the preictal one (plot (a)).

This impression is confirmed by a statistical analysis: The mean vectors of segment length differ significantly between the preictal and ictal period (defined as above), as an $F$-test at a significance level of 0.99 confirms. In detail, the $F$-test is constructed as follows:

Let $\mu_i = (\mu_{A1},\ldots,\mu_{C5})_i^T$ with $i = 1$ symbolizing preictal and $i = 2$ ictal and e.g. $\mu_{A1}$ denoting the mean segment length of channel A1 in the respective period $i$ as depicted in Fig. 8.4.1. We test $\mathcal{H}_0 : \mu_1 = \mu_2$ vs. $\mathcal{H}_1 : \mu_1 \neq \mu_2$. $\mathcal{H}_0$ is rejected if the test statistic exceeds the quantile (Anderson 2003),

$$\frac{N_1+N_2-K-1}{(N_1+N_2-2)K}T^2 > F(K, N_1+N_2-K-1; 1-\alpha). \qquad (8.4.1)$$

Hereby $N_1$ and $N_2$ denote the respective sample size and $T^2 = \frac{N_1 N_2}{N_1+N_2}(\bar{\mathbf{x}}_1 - \bar{\mathbf{x}}_2)^T \hat{\mathbf{S}}^{-1}(\bar{\mathbf{x}}_1 - \bar{\mathbf{x}}_2)$, where $\bar{\mathbf{x}}_i$ denotes the respective empirical mean and $\hat{\mathbf{S}}$ the pooled empirical covariance matrix[7]. $F$ is the quantile of the F-distribution. For $\alpha = 0.01$ we have $5.7 > 2.7$, i.e. we reject $\mathcal{H}_0$. In other words, segment lengths differ significantly between the ictal rhythmic $\vartheta$-activity of seizure 1 and the preceding period.

Second, we briefly want to discuss the choice of the threshold. As Fig. 8.4.2 details in the form of box plots, segment lengths grow with increasing threshold both preictally (plot (a)) and ictally (plot (b)).[8] Due to the low threshold th $= 0.07$ we obtain a reactive segmentation behavior and short segments. This choice turned out to deliver accurate results for the analysis of seizure propagation, as will be discussed in Subsection 8.4.3. In particular the influence of the initial reference point $m^*$ is almost negligible in this setting.

---

[7] The pooled empirical covariance matrix is $\hat{\mathbf{S}} = \frac{1}{N_1+N_2-2}\left[(N_1-1)\hat{\mathbf{S}}_1 + (N_2-1)\hat{\mathbf{S}}_2\right]$, with $\hat{\mathbf{S}}_i$ the estimated covariance matrix of the respective sample.

[8] Intervals defined as before: 1 min prior to the onset of rhythmic $\vartheta$-activity (at 16:12:45) and 1 min of this rhythmic activity.

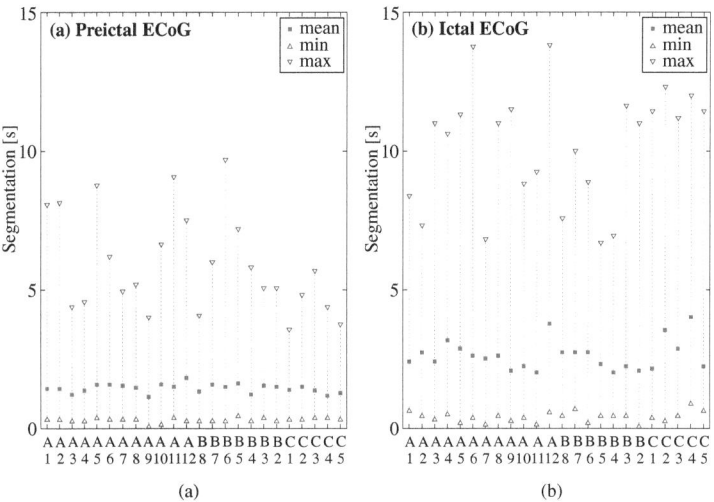

**Figure 8.4.1:** *Segmentation behavior preictally and ictally.* Analysis for seizure 1. (a) 1 min preictal (16:11:45 - 16:12:45), (b) 1 min ictal rhythmic $\vartheta$-activity (16:12:45 - 16:13:45). Segments are longer in the ictal rhythmic period than in the preictal one.

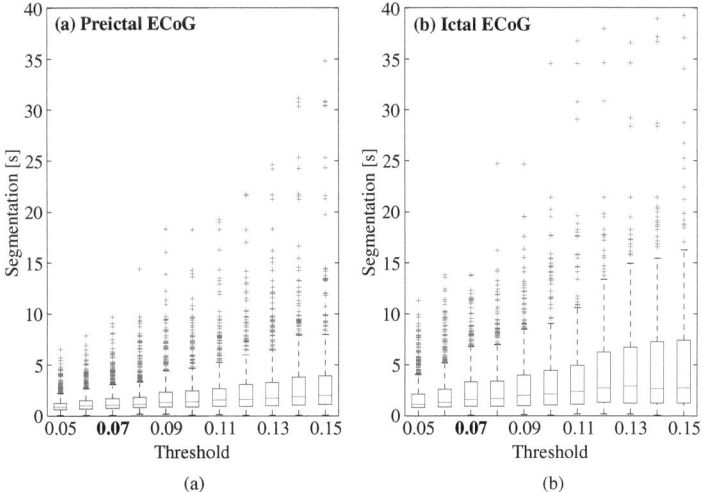

**Figure 8.4.2:** *Influence of threshold on segment length.* Analysis for seizure 1 in form of box plots. (a) 1 min preictal (16:11:45 - 16:12:45), (b) 1 min ictal rhythmic $\vartheta$-activity (16:12:45 - 16:13:45). Segment length grows with increasing threshold in both cases.

## 8.4.2 Classification

As mentioned in Subsection 8.2.3, the classification step of our methodology is clinically inspired. It imitates the visual ECoG analysis performed by clinicians at the Neurological Center Rosenhügel. Therefore, we think that the proposed algorithm (first: distinct epileptic activity; second: initial epileptic activity) is reasonable. Note that high rhythmicity is a necessary criterion in all three classification steps in order to avoid false-positives (e.g. high-amplitude artifacts with arhythmic character must not be classified as epileptic).

In our opinion the proposed rule set reflects the statistical properties of epileptic activity well, compare the illustration in Fig. 8.3.3. However, possible ameliorations are conceivable: For instance, improved classification rules could additionally consider the entropy as measure of rhythmicity (van Putten et al. 2005) or classify the waveforms in the time domain (Fürbaß et al. 2012).

## 8.4.3 Seizure propagation

In this pilot study we assume the initial spread of $\vartheta$-activity to be a valid indicator (among others) for seizure propagation in focal epilepsy, in particular for the determination of the SOZ (Foldvary et al. 2001). Our results are in good accordance with the clinical findings (see Table 8.3.1), which supports our assumption.

As the visual analysis reveals, data quality declines in the course of the seizures (compare Fig. 8.3.4). This is reflected by the visual analysis of the three clinical experts: In seizures 1 and 2 the latency between the indicated start of epileptic activity (by each of the three experts) on each channel lies between 500 ms and 1 s (indications of expert 1 leading, followed by experts 2 and/or 3). In seizure 3 these latency times increase up to 2 s which is due to the decreased data quality (on some channels it is difficult to determine the onset of epileptic activity visually).

Consequently, the quality of our findings is better in seizures 1 and 2 than in seizure 3 as well. In the first two seizures the epileptic starting points indicated by our algorithm (on each channel) match the visual analysis better than in the third seizure, see Fig. 8.3.4. This is also reflected in the analysis of initial electrodes as summarized in Table 8.3.1.

Due to the decreasing data quality in seizure 3 we propose to put more emphasis on the first two seizures in the SOZ analysis. This would restrict the SOZ to the parieto-occipital region of electrodes B8, A12 and A11, which is the posterior part of the area discussed in Subsection 8.3.4 and marked in Fig. 8.3.5.

A validation of the determined SOZ, in particular of the restricted one, could be achieved by means of post-surgical outcome as detailed in Subsection 2.3.6. However, the surgical intervention in the patient is still pending at the moment of publication, compare Chapter 10.

We finally want to mention that the combination of a simple segmentation method and a clinically motivated classification step delivers results which are consistent and well correlated with clinical findings. In our opinion this is due to the close relation between the method and neurophysiology.

### 8.4.4 Concluding remarks

In this chapter we proposed a novel method for early seizure propagation analysis based on segmentation and subsequent classification of ictal ECoG data. The methodology allows to determine the initial seizure spread (in particular the SOZ) and yields results which are well correlated with the visual analysis of clinicians. It therefore has the potential for an objectivation in the presurgical clinical evaluation of therapy-resistant patients.

However, this requires further research: Next steps include the application to a broader data basis (at least 5 patients undergoing invasive long-term monitoring) and the evaluation of the algorithm's performance (sensitivity vs. specificity, mean latency time of detected epileptic activity). Post-surgical outcome of these patients would serve as an additional performance indicator.

# Part III

# Results and Discussion

# 9 Framework results

Χρὴ δὲ τὸν μέλλοντα ὀρθῶς προγιγνώσκειν τοὺς περιεσομένους καὶ τοὺς ἀποθανουμένους, ὅσοισί τε ἂν μέλλῃ πλέονας ἡμέρας παραμένειν τὸ νούσημα καὶ ὅσοισιν ἂν ἐλάσσους, τὰ σημεῖα ἐκμανθάνοντα πάντα δύνασθαι κρίνειν, λογιζόμενον τὰς δυνάμιας αὐτέων πρὸς ἀλλήλας, ὥσπερ διαγέγραπται περί τε τῶν ἄλλων καὶ τῶν οὔρων καὶ τῶν πτυέλων, ὅταν ὁμοῦ πῦόν τε ἀναβήσσῃ καὶ χολήν.

— Ἱπποκράτης: *Προγνωστικόν*

He who would correctly beforehand those that will recover, and those that will die, and in what cases the disease will be protracted for many days, and in what cases for a shorter time, must be able to form a judgment from having made himself acquainted with all the symptoms, and estimating their powers in comparison with one another, as has been described, with regard to the others, and the urine and sputa, as when the patient coughs up pus and bile together.

— Hippocrates, *The Book of Prognostics*[1]

## 9.1 Introduction

In this chapter we are concerned with the evaluation of the framework for epileptic seizure propagation analysis proposed in Chapter 4. For this purpose we compile the results of the individual methods from Chapters 5 (HFO propagation), 6 (Causality analysis), 7 (Influence Analysis), 8 (Segmentation) and compare them to the visual analysis by clinical experts from Section 4.3.

The seizure analysis (see Section 9.3) will be supplemented by complementary clinical findings from Section 9.2. This holistic approach allows

---

[1] Opening of part 25; English translation by Francis Adams.
Original text and English translation are accessible online via the PERSEUS project at www.perseus.tufts.edu (»prognosticon«).

us to deduce indications for the SOZ and early seizure spread (see Section 9.4).

## 9.2 Previous clinical findings

First, we consider supplementary clinical findings which were available at the time of the invasive long-term video-EEG monitoring in 2011.[2] The anamnestic report comprises findings of the following five diagnostic methods:

**MRI:** Two MRI scans were performed, the first in May 2005 and the second in March 2007 at DiagnoseZentrumUrania, Vienna (3 T). Both scans were without pathological findings.

**PET:** The results of two PET scans are available. A first PET scan was acquired at Wilhelminenspital Vienna in June 2009 (202 MBq F-18-FDG) and revealed an increased glucose metabolism in the left gyrus frontalis inferior. However, a second PET scan at Vienna General Hospital 2 months after the invasive video-EEG monitoring, i.e. in January 2012 (708 MBq C-11-methinione, 266 MBq F-18-FDG), was without clear pathological findings.

**Neuropsychological testing:** The patient underwent neuropsychological tests at Neurological Center Rosenhügel, Vienna, in October 2009 and in November 2011. Results were as follows: right-handed patient with an average general intellectual level (IQ: 94), deficits in selective attention, divergent thinking (phonematic verbal fluency) and in the memory span of the short term and working memory (visual and verbal).

**fMRI:** A functional MRI scan was done at Vienna General Hospital in September 2009 in order to determine the language lateralization (Desmond et al. 1995). It revealed a left-lateralized response specific

---

[2]Compare Subsection 2.3.3 for the diagnosis in epilepsy patients.

to speech by language-associated activation of Broca's and Wernicke's areas.

**Surface EEG:** Prior to the invasive long-term video-EEG monitoring the patient underwent a surface EEG monitoring at Neurological Center Rosenhügel in April 2009. At that time, three seizures were recorded, but no exact localization of the SOZ could be derived from the ictal EEG (compare the discussion in Subsection 10.2.1). The interictal EEG revealed series of spikes (duration of one to two seconds) with a maximum at P8, postictally regional spikes with a maximum at FT10 and T8 and right-hemispheric slowing.

Symptomatology of the three seizures was as follows: pausing, version of the head to the left (contralateral sign), cloni of the left corner of the mouth (contralateral sign), »figure of four« sign with extension of the left and flexion of the right upper extremity (extension as contralateral, flexion as ipsilateral sign) and asymmetric termination of the seizure with cloni of the right upper extremity (ipsilateral sign). Postictal testing did not reveal any paresis or aphasia.

These findings point to an MRI-negative right-hemispheric focal epilepsy, although the surface EEG itself does not allow for a lateralization. The reasoning is as follows: First, we observe intermittent (postictal) right-hemispheric slowing. Second, the clinical symptomatology (ipsi- and contralateral signs, see e.g. Rosenow and Lüders (2001)) indicates a right-hemispheric onset. Finally, the postictal lack of aphasia suggests that the dominant language hemisphere (left according to fMRI) is not initially involved in the seizures.

In contrast, the findings of the neuropsychological tests are not conclusive. Impaired phonematic verbal fluency and deficits in the short term and working memory would imply a frontal functional deficit zone[3], the deficits in the visual memory a right temporal deficit zone. However, during or shortly before both of the test sessions the patient suffered from seizures which might lead to a bias of the test results.

---

[3]Compare Subsection 2.3.5 for the concept of the cortical zones.

Note that the diagnostic imaging (MRI for localization of a lesion, PET for additional information regarding the SOZ) did not provide any additional information.

## 9.3 Analysis of seizures

As was pointed out in the last section, surface EEG recordings from 2009 provide an insufficient localizing value. In order to confirm the preliminary diagnosis of right-hemispheric focal epilepsy and exactly localize the SOZ, the patient underwent an invasive long-term video-EEG monitoring in November 2011. We refer to Section 4.3 for the complete anamnesis.

Here we are concerned with the analysis of the invasively recorded seizures in 2011, compare Figs. 4.3.2, 4.3.3 and 4.3.4 for the recordings.

First, we consider the ictal ECoG. The results of the seizure propagation analysis are summarized in Table 9.3.1. It details the initial electrodes and follow-up electrodes as determined by each of the four technical methods (HFO detection according to Chapter 5, causality analysis according to Chapter 6, influence analysis according to Chapter 7, segmentation according to Chapter 8). The results of the visual inspection of HFOs and rhythmic $\vartheta$-activity are listed in Table 4.3.1.

Second, we additionally examine the interictal ECoG. It reveals frequent paroxysmal fast activity at electrodes A5 to A12, B8 and C1 to C5. Furthermore spikes with a maximum at A3 and independently spikes with a maximum at A9, A11 and C4 can be observed.

Table 9.3.2 condenses these findings. It lists the respective SOZ and early seizure spread as determined by each of the four technical methods and by the visual analysis. These results suggest a temporo-parieto-occipital SOZ on the right hemisphere with an early seizure spread in frontal (strip A) and mesial (strip C) direction, compare Fig. 9.3.1 for a visualization based on the patient's MRI scan. In this graphical representation, each of the four inferred SOZs is highlighted in different color, electrodes be-

| # | Propagation | HFO det. | Causality | Influence | Segmentation |
|---|---|---|---|---|---|
| 1 | onset | B8, B7 | A7, A10, A11, C2, C5 | A10, A12 | B8 |
| 1 | follow-up | A11, A4 | - | - | A10, A11, A12 |
| 2 | onset | A9 | A10 | A5, A6 | A11 |
| 2 | follow-up | - | - | - | A12, B7 |
| 3 | onset | - | A10 | A6 | A7 |
| 3 | follow-up | - | - | - | A1, A9, A10, C4, C5 |

**Table 9.3.1:** *Detailed analysis of the three seizures.* Initial and follow-up electrodes according to the 4 technical methods.

| | Method | SOZ | Early seizure spread |
|---|---|---|---|
| Technical methods | HFO detection | A9, B7 - B8 | A4, A11 |
| | Causality analysis | A10 | - |
| | Influence analysis | A5 - A12 | - |
| | Segmentation | A11 - A12, B8 | A1, A9 - A10, B7, C4 - C5 |
| Medical report (visual inspection) | | Temporal caudal (A9 - A12) | A strip, B7, C strip |

**Table 9.3.2:** *Seizure propagation according to seizure analysis.* SOZ and early seizure spread according to the 4 technical methods and the visual inspection.

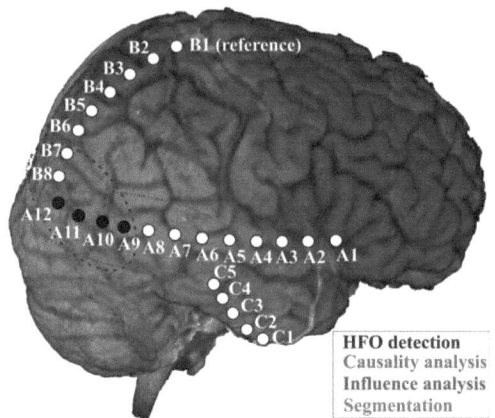

**Figure 9.3.1:** *SOZ according to seizure analysis.* Respective SOZ according to each of the 4 technical methods and the visual inspection is indicated. SOZ of technical methods is indicated by ovals in color; HFO detection: magenta, Causality analysis: green, Influence analysis: red, Segmentation: blue. Electrodes belonging to the area indicated in the medical report are marked in black.

longing to the area indicated by the medical report are marked in black. As can be seen from this representation, the different technical results are in good accordance with each other as well as with the visual inspection. In particular we observe a distinct overlap of the five indications in the temporo-parieto-occipital region, compare the discussion in Subsection 10.2.2.

## 9.4 Overall findings

In the final step we derive a common SOZ and the direction of initial seizure spread. As suggested by the framework for epileptic seizure propagation analysis in Chapter 4, we consolidate the results of the seizure analysis (technical methods as well as visual inspection according to Section 9.3) and the supplementary clinical findings (see Section 9.2).

Based on these findings we conclude on an MRI-negative right-hemispheric focal epilepsy with a temporo-parieto-occipital SOZ involving the electrodes A9-A12, B8. These five electrodes represent the area with

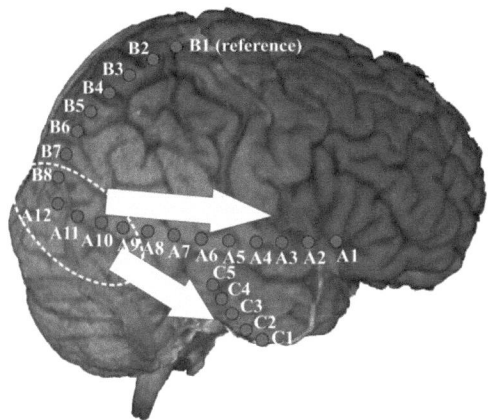

**Figure 9.4.1:** *Seizure propagation according to framework.* SOZ comprises the temporo-parieto-occiptial area of electrodes A9-12, initial seizure spread in frontal (A strip) and mesial direction (C strip).

substantial overlap of the five different indicated SOZs in Fig. 9.3.1. Furthermore, early seizure spread propagates in frontal (strip A) and mesial (strip C) directions. Compare Fig. 9.4.1 for an illustration of early seizure propagation.

These findings, in particular the SOZ, require validation by post-surgical follow-up (seizure freedom according to Engel (1996b), compare Table 2.3.3), as will be discussed in Section 10.1.

# 10 Discussion and Outlook

> Le but de la discussion ne doit pas être la victoire, mais l'amélioration.
>
> — Joseph Joubert: *Pensées*
>
> The aim of discussion should not be victory, but progress.
>
> — *Thoughts*[1] by Joseph Joubert (1754-1824), French moralist

## 10.1 Patient

We want to start the discussion of the framework results with aspects related to the patient analyzed in this study.

The patient was admitted in the case of suspected MRI-negative right-hemispheric focal epilepsy for invasive video-EEG monitoring in November 2011. The neurophysiologists expected initial ictal activity in the temporal region, and the subdural electrodes were implanted accordingly at the General Hospital of Vienna, University Clinic of Neurosurgery (see Fig. 4.3.1). Therefore, while the entire temporal region (anterior and posterior) of the patient's brain was appropriately covered by subdural strip electrodes, the occipital region of the patient's brain was not.

Surprisingly, the invasive video-EEG monitoring revealed initial ictal activity in the temporo-parieto-occipital area of electrodes A9-A12, not in the anterior temporal region. Due to the low coverage of the occipital region it was impossible to decide whether electrodes A9-A12 indicated the SOZ, or whether they recorded propagated activity from a posterior origin in fact. In the latter case, the SOZ would be located somewhere in the occipital, not in the temporal lobe.

---

[1]Own translation.

In order to answer this question and to locate the SOZ reliably, the patient was re-admitted for intracerebral long-term video-EEG monitoring in May 2012. For this purpose, nine depth electrodes with a total of 72 contact points were implanted at the General Hospital of Vienna, University Clinic of Neurosurgery. Fig. 10.1.1 shows two MRI scans with the electrode positions. As can be seen from these scans, all relevant brain structures of the entire right hemisphere are covered by electrode positions.

Recording was done at a sampling frequency of 1024 Hz (as for the ECoG monitoring in November 2011), and recordings were referenced to contact point B12 (most anterior) in order to avoid referencing to any region showing initial ictal activity. The numbering of the respective electrode contact points starts at the electrode end in the inner brain structures, i.e. the contact point with the highest number is near the brain surface.

During the recording period of one week, the patient suffered from five seizures. Unfortunately, the depth electrode recordings of these seizures did not allow for a localization of the SOZ. In each of the five seizures, ictal activity starts nearly simultaneously in the entire right hemisphere except of the frontal region. Fig. 10.1.2 shows the depth electrode recordings for seizure 5, and Table 10.1.1 summarizes the ictal onset times on the respective electrodes according to the visual inspection. At 08:42:31 we observe an onset of ictal activity on electrodes D and G, i.e. in occipital and temporal brain structures (compare Fig. 10.1.1, plot (b)). One second later, at 08:42:32, ictal activity starts at contact points A7-A10 (posterior end of depth electrode A) and on electrodes F and I, i.e. again in both occipital and temporal brain structures. Moreover, onset on electrode C (located between electrodes D and F, see 10.1.1, plot (b)) is only observed another second later at 08:42:33. This observation is somehow contradictory to any anatomically reasonable propagation pattern.[2]

As the patient was suffering from an increased seizure frequency and

---

[2]Note that we do not observe any activity in the frontal lobe which corresponds at least to the situation expected: Only contact points B1-B3 (inner locations near electrode H and the amygdala) of the anterior electrode B are affected.

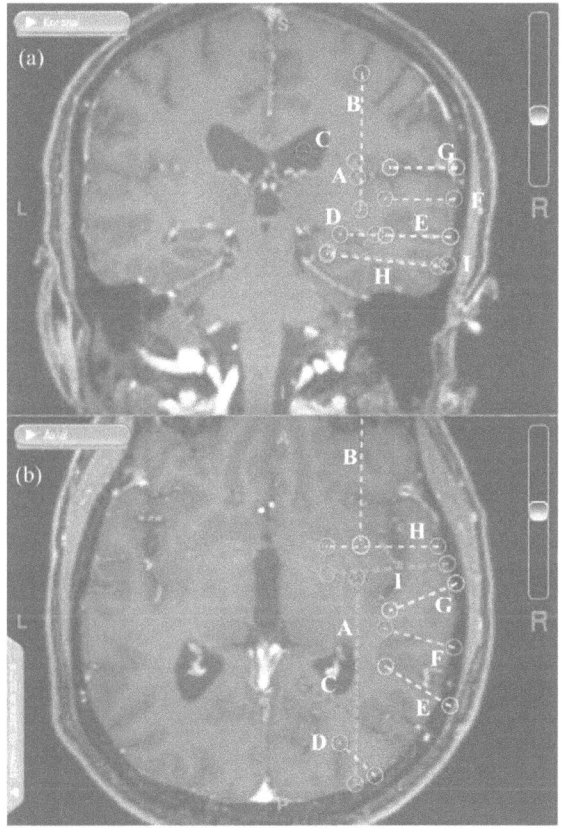

**Figure 10.1.1:** *Depth electrode positions.* MRI scans in (a) coronal view, (b) axial view. Scan orientation is indicated: R - right, L - left, P - posterior, A - anterior, I - inferior, S - superior. Targets of electrodes are as follows: A (green) - posterior insula, B (violet) - anterior insula, C (violet) - cingulum, D (violet) - occipital, E (blue) - T2 posterior, F (red) - T1 retro-insular, G (orange) - parietal operculum, H (red) - amygdala temporal, I (green) - hippocampal head.

subjective low quality of life (compare Subsection 2.3.7), a surgical intervention still seemed desirable. Thus, as another attempt to SOZ determination, the patient underwent an ictal SPECT at the General Hospital of Vienna, University Clinic of Nuclear Medicine, in January 2013. However, diagnostic imaging only revealed expended ictal activity within the right hemisphere, no further refinement was possible.

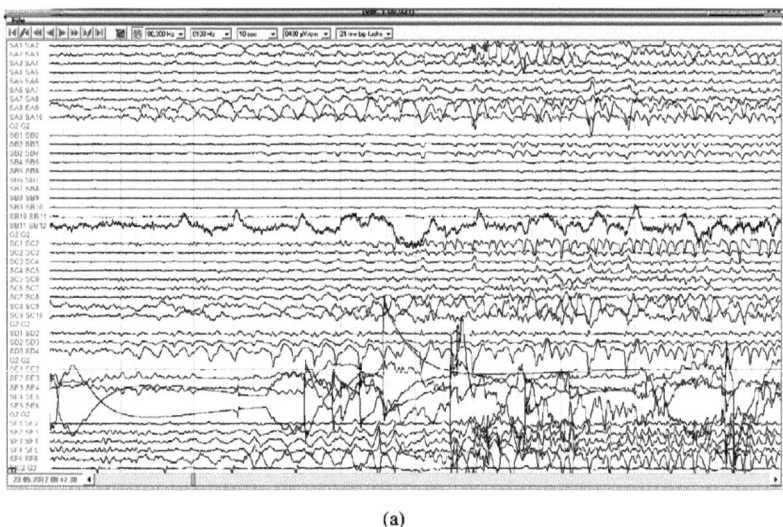

(a)

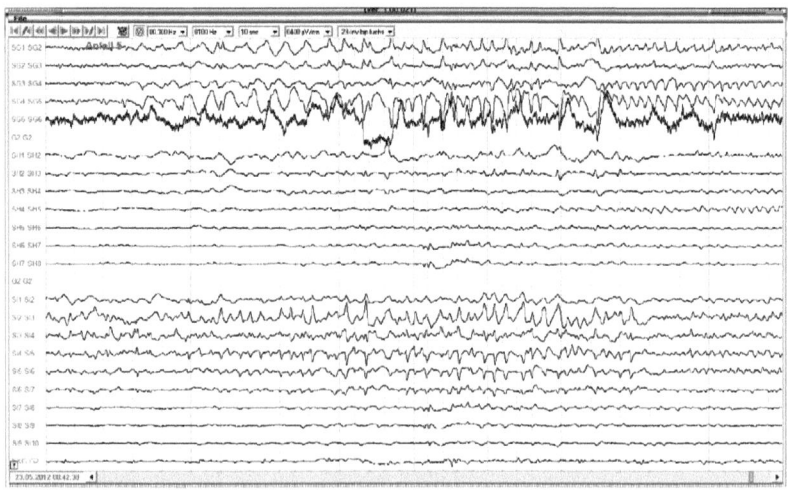

(b)

**Figure 10.1.2:** *Intracerebral EEG.* Initial 10 seconds of exemplary seizure 5 in bipolar setup. (a) depth electrodes A - F, (b) depth electrodes G - I.

| Depth electrode | Contact points | Onset time | Affected contacts points |
|---|---|---|---|
| A | 10 | 08:42:32 | A7-A10 |
| B | 12 | 08:42:35 | B1-B3 |
| C | 10 | 08:42:33 | C7-C10 |
| D | 4 | 08:42:31 | D2-D4 |
| E | 6 | 08:42:33 | E1-E6 |
| F | 6 | 08:42:32 | F3-F6 |
| G | 6 | 08:42:31 | G1-G6 |
| H | 8 | 08:42:38 | H3-H4 |
| I | 10 | 08:42:32 | I2-I6 |

**Table 10.1.1:** *Seizure onset in intracerebral recordings in exemplary seizure 5. Onset of ictal activity on different depth electrodes happens with minimal delay only.*

Therefore, a surgical intervention was still pending at the moment of completion of this study. The decision becomes more complicated by the fact that a resection in the occipital region has a direct impact on the visual cortex and might lead to (partial) loss of sight. This risk and the potential chance of seizure freedom have to be carefully weighed against each other by both the patient and the involved clinicians.

## 10.2 Framework

This section is dedicated to the discussion of the seizure propagation framework itself. We will consider two aspects, the limitation of the methodology to invasive data in Subsection 10.2.1 and the quality of our findings in Subsection 10.2.2.

### 10.2.1 Limitation to ECoG data

From a practical point of view, the analysis of surface EEG has a number of advantages over invasive techniques: surface EEG recordings are easy and cheap to perform and sterility is not required due to the non-invasive character of the examination, compare Subsection 2.2.1. Consequently, the patient's risks are significantly reduced (no surgical intervention for electrode implantation requiring specialized know-how, no risk of infections). However, the proposed methodology for the exact localization of the SOZ is limited to invasive EEG (in particular ECoG in this study) for three reasons.

First, the quality of surface EEG recordings is often impaired by artifacts, compare Subsection 2.2.4. In the context of a presurgical examination such artifacts make the visual inspection of the recordings difficult or even render an analysis impossible. For instance, in the tonic-clonic phase of a seizure muscle artifacts typically blur the picture. In the worst case, electrodes become loose during motor seizures, and data cannot be acquired reliably any longer.

In particular automated EEG-analysis techniques require recordings with no or only few artifacts. As a prominent example, ocular artifacts severely disturb algorithms due to their large amplitude. Therefore, various technical methods for automatic removal of artifacts have been proposed in literature, with increasing complexity in the last years. They can be divided into regression-based and component-based approaches on a coarse level (Wallstrom et al. 2004), see Table 10.2.1 for an overview of various approaches for eye and muscle artifact correction.

However, the application of such artifact-removal algorithms is not straight-forward and might lead to induced filtering artifacts. As an example, consider Fig. 10.2.1. Here, we exemplarily apply the methodology of Schlögl et al. (2007) to the first seizure recorded in the course of the video-EEG monitoring in 2009. Plot (a) depicts the original surface EEG recordings in reference setup, plot (b) the corrected data. In both visualizations, the fronto-polar channels Fp1 and Fp2 are highlighted in red for better orientation. In plot (a), the potential induced by the eye movement does not only impair the fronto-polar channels, but also the frontal and temporo-parietal ones (which we are interested in). In plot (b), these artifacts are removed by the algorithm. However, while channels F7, TP9 and TP10 are appropriately corrected, channel P7 is impaired by the correction.

Therefore, we refrain from the analysis of automatically corrected sur-

| Artifact | Approach | Method | References |
|---|---|---|---|
| Eye | Regression-based | Adaptive filtering | He et al. (2004), Puthusserypady and Ratnarajah (2006) |
| | | Regression | Schlögl et al. (2007) |
| | Component-based | ICA | Vigario (1997), Jung et al. (1998), Jung et al. (2000), Delorme et al. (2007), Mantini et al. (2007); temporally constrained: James and Gibson (2003); Bayesian classification: LeVan et al. (2006); time-variant: Boudet et al. (2007); blind source separation: Joyce et al. (2004), Hallez et al. (2009); FASTER algorithm: Nolan et al. (2010) |
| | | PCA | Liu and Yao (2006) |
| Muscle | Spectrum-based | Regression | Gasser et al. (2005) |

**Table 10.2.1:** *Recent automatic artifact removal approaches.* Overview of methods for eye and muscle artifact correction in EEG.

face EEG recordings and stick to invasive data, which are not impaired by ocular (and mostly muscle) artifacts.

Second, in Chapter 5 we analyze the initial propagation of ictal HFOs as a highly specific bio-marker of the SOZ. Due to the character of HFOs (high-frequency and low-amplitude) these EEG correlates are mostly filtered by the cranium, compare Subsection 2.3.5. Therefore, it is difficult to perform HFO propagation analysis in surface EEG recordings, independently of the methodology (visual analysis or automatic detection). Although HFOs have been detected in surface EEG recordings of children (Wu et al. 2008, Inoue et al. 2008, Kobayashi et al. 2009) and very recently in adults (Andrade-Valenca et al. 2011), we assume that these studies represent rare cases under favorable circumstances. As we aim at an HFO analysis during presurgical work-up in as many patients as possible, we see the need to stay with

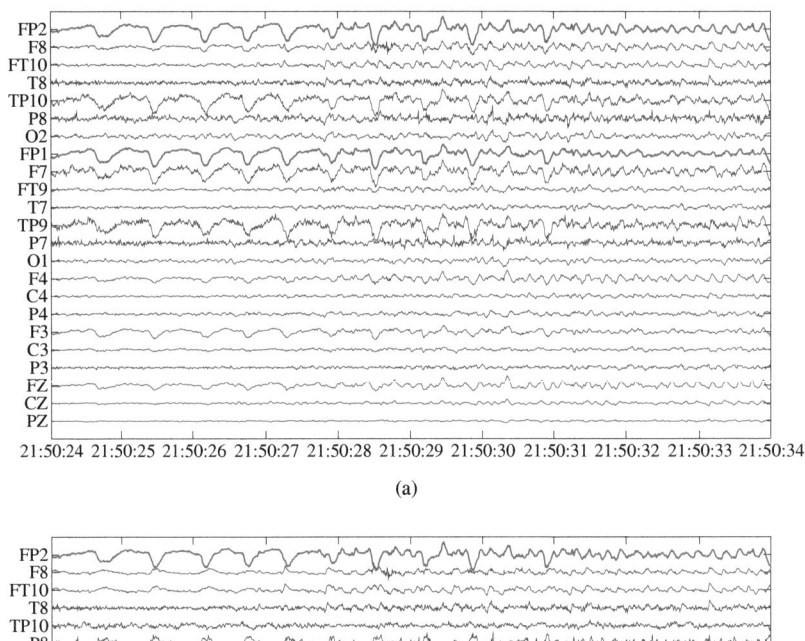

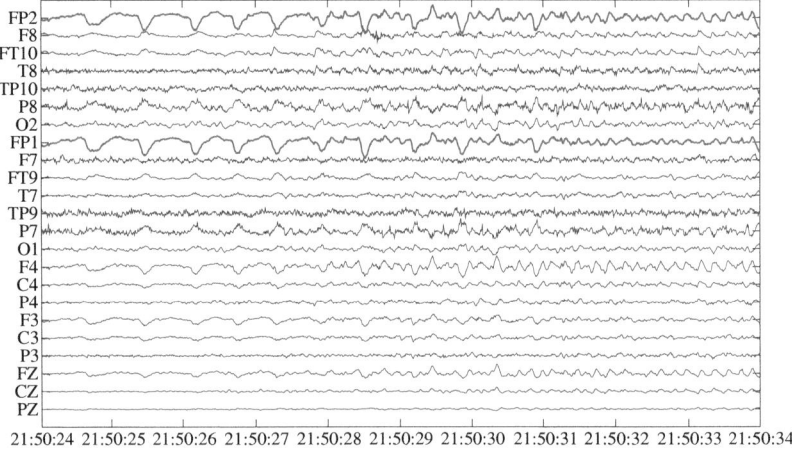

**Figure 10.2.1:** *EOG artifact correction.* Exemplary surface EEG recordings in reference setup from the long-term video-EEG monitoring in April 2009 (seizure 1). Seizure onset on the right hemisphere at 21:50:26. (a) original data, and (b) correct data, fronto-polar channels Fp1 and Fp2 highlighted in red. Channels F7, TP9 and TP10 are appropriately corrected, but P7 is impaired by correction.

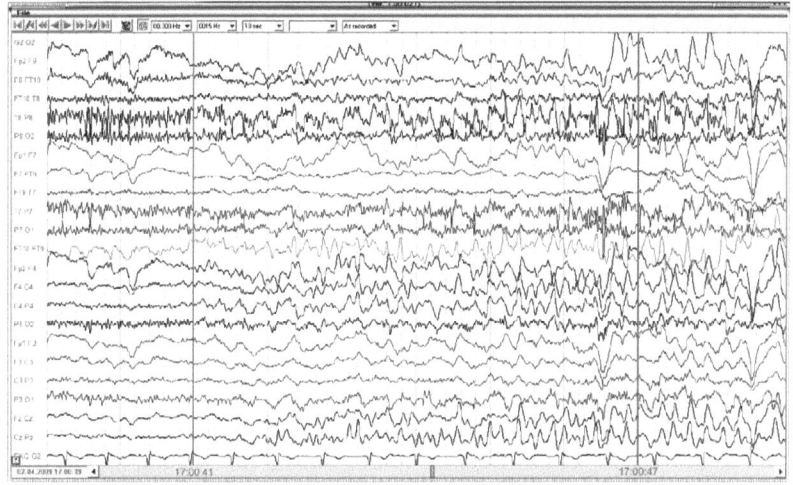

**Figure 10.2.2**: *Low spatial resolution of surface EEG. Exemplary surface EEG recordings in bipolar longitudinal setup from the long-term video-EEG monitoring in April 2009 (seizure 3). Seizure onset on the whole right hemisphere at 17:00:41 and generalization at 17:00:47 (both events indicated by vertical orange lines).*

invasive recordings.

Finally, the major argument for usage of invasive data is spatial resolution. First of all, invasive EEG allows for a localization of the SOZ on a higher level (focal vs. regional) by definition, see Subsection 2.2.5. Furthermore, surface EEG sometimes even does not allow for the localization of the SOZ on a regional level due to rapid spread of ictal activity. As an example, consider the surface EEG recordings from 2009. They only allowed for determination of the seizure onset on a hemispheric level, as mentioned previously in Section 9.2. Fig. 10.2.2 shows seizure 3 of these recordings in bipolar longitudinal setup. According to the visual inspection performed by clinicians, onset took place on the whole right hemisphere at 17:00:41 (marked in orange), followed by secondary generalization at 17:00:47 (marked in orange). Note that for better visibility the recordings are low-pass filtered at 15 Hz in order to suppress muscle artifacts[3].

---

[3] Compare the discussion about artifact suppression above.

## 10.2.2 Performance

In Chapter 9 we showed that the findings derived from the framework for epileptic seizure propagation analysis correlate well with the medical report based on visual inspection. Overall, we have a conclusive result on seizure initiation and propagation in the analyzed patient (see Fig. 9.4.1). However, the individual performance of the respective technical methods shows a considerable variation, compare Table 9.3.1. Hereby, the quality of our findings depends on the character of the method:

On the one hand, it is remarkable that the methods which are closest to neurophysiology in design deliver results which are best in accordance with the visual inspection. In particular, the segmentation method proposed in Chapter 8 is strongly inspired by clinical reasoning and therefore yields excellent results. Moreover, the automatic detection of ictal HFOs proposed in Chapter 5 performs very well due to its close relation to neurophysiological frequency bands (in the pre-emphasis as well as in the detection step).

On the other hand, the performance of the causality-based methods (causality analysis in Chapter 6, influence analysis in Chapter 7) is varying. While the underlying idea of tracking hyper-synchronous activity is physiologically inspired, the methodology itself is purely theoretical in nature. The concept of Granger causality and the derived concept of influence are based on mathematical assumptions in a theoretical setting, which are not necessarily satisfied in application to ECoG data.[4] Moreover, Granger causality is only one possible way of defining the concept of causality in a formal way (Pearl 2000) and might not be the optimal modeling for ictal dependencies in ECoG. Finally, the employed methods do not explicitly differentiate between ictal synchronous activity and physiological couplings and feedback processes, compare the discussion in Subsection 6.4.3. This discrepancy between theory and application leads to the observed variation in quality of our findings reported in Chapters 6 and 7.

A general problem of any kind of quantitative as well as visual EEG

---

[4]Compare Section 3.3 for the theoretical background of Granger causality.

analysis is *coverage*. While cortical activity in only measured at certain points defined by the electrodes, the majority of the neurophysiological processes is not accessible to the analysis. In particular, subdural strip electrodes only track neuronal activity on the surface of the brain. However, the focus of a seizure is usually located in deeper brain regions or is not directly covered by an electrode. Therefore, the employed quantitative methodology as well as the visual analysis can only indicate a circumscribed area where the epileptic activity is noticed first. However, this is not necessarily the SOZ. This is a fundamental issue in case of the analyzed patient, compare the discussion in Section 10.1.

In order to circumvent this limitation, efforts have been made in the last years to increase the number of electrodes. Ostenveld and Praamstra (2001) proposed a new positioning system in surface EEG which defines the positions of 128 electrodes on the skull (even up to 345 positions supported). This so-called *high-density EEG* has become important in 3D source localization, as numerous studies have proved an increased localization of the epileptic focus in this setting (Lantz et al. 2003, Michel et al. 2004).

In invasive EEG, the use of depth electrodes is a resort. Their individual positioning allows to capture intracerebral activity, but the implantation demands a laborious planing in order to avoid vascular damage. Cortical recording is an option with less effort. Here, subdural grid electrodes provide a much better coverage than subdural strip electrodes. However, their implantation requires a larger craniotomy due to their spacious layout with an increased risk of severe complications, e.g. bleeding or infections (Van Gompel et al. 2008).

We want to end this discussion with a general remark on performance issues of signal processing methods in ECoG analysis.

Throughout this study we employ well-known techniques which are established in signal processing, e.g. signal detection (Chapter 5) or non-parametric spectral estimation (Chapter 8). Despite their wide popularity in applications like radar analysis or telecommunications the performance of the algorithms in ECoG analysis is variable and requires further tuning prior to a roll-out to a larger patient group. Here we briefly want to discuss possible reasons for this discrepancy.

In classical engineering applications we have an excellent modeling of the underlying physics at hand. Take mobile telecommunications as an example. In this case, the characteristics of both sender and receiver are specified by design, only the information transmission in between has to be modeled. This step is based on physics of electromagnetic wave transmission which is described very well by the Maxwell equations (Jackson 1998). In order to further increase the algorithm performance, electro-engineers have the possibility to send test signals for parameter calibration.

The situation is different in case of quantitative approaches in neurophysiology, where each patient is individual. In any kind of quantitative EEG analysis, one aims at modeling at a macroscopic level. As the system characteristics itself are not known at that level, we can only measure outputs of this black-box model and employ data-driven approaches. On a microscopic level, however, the bio-chemical mechanisms are well understood (Kandel et al. 2012), and neuronal modeling approaches have been successful for more than 50 years (Hodgkin and Huxley 1952). We therefor lack a continuous description from microscopic to macroscopic level for successful signal processing in ECoG. There has been increasing interest in *multi-scale modeling* with significant progress recently (Demongeot et al. 2003). In particular, Eliasmith et al. (2012) successfully simulated a 2.5-million neuron model very recently. Note, however, that this enormous complexity is still far away from the one of the human brain.

## 10.3 Outlook

In this study we proposed a novel framework for epileptic seizure propagation analysis in ECoG recordings of therapy-resistant patients suffering from focal epilepsy. The findings derived from our methodology are in very good accordance with the clinical findings based on visual inspection of the raw data. Therefore, we are confident that this framework has the potential to determine the SOZ and the initial seizure spread in the course of a presurgical examination, thus supporting clinicians by objectivating their findings.

However, further research has to be conducted for this purpose.

First, the specific results of this study need to be evaluated by a post-surgical follow-up of the patient. A post-surgical outcome of Engel class I would serve as confirmation of our findings.

Second, the technical methods of the framework have to be refined, as discussed in the respective chapters. Furthermore, slight adaptations of the methodology might allow for an application to intracerebral recordings.

Finally, the framework has to be validated by application to a larger number of patients. This important step involves an extensive post-surgical follow-up of the patients (post-operative controls 4 months, 1 year, 2 years and 5 years after the intervention). Seizure controls together with an analysis of post-surgical MRI scans and neuropsychological tests allow for a thorough statistical assessment of the performance of the framework as well as of the individual methods. In particular this step will help to understand under which circumstances (e.g. syndrome, seizure type, propagation speed) the proposed methodology yields reliable results.

# Part IV

# Appendix

# A Proofs for the statistical background

## A.1 Non-parametric spectral estimation

### A.1.1 Preliminary definitions

In this subsection we will prove the representations of the discrete-time Fourier transform (DTFT) of the centered and non-centered rectangular window.

First, we prove Lemma 1.

*Proof.* DTFT of the centered rectangular window.

Immediate by using the definition of the DTFT, i.e.

$$\begin{aligned}
D_N(f) = \mathfrak{F}\left\{w_N^{cR}[n]\right\} &= T \sum_{n=-\infty}^{\infty} w_N^{cR}[n] e^{-2i\pi fTn} \\
&= T \sum_{n=-\frac{N}{2}}^{\frac{N}{2}} e^{-2i\pi fTn} \\
&= T \frac{1 - e^{-i\pi fTN}}{1 - e^{-i\pi fT}} \\
&= T \frac{\sin(N\pi fT)}{\sin(\pi fT)}.
\end{aligned}$$

$\square$

This allows us to derive the result of Corrolary 1.

*Proof.* DTFT of the rectangular window.

The DTFT allows for time-shifting, i.e.

$$\mathfrak{F}\left\{x[n-n_d]\right\} = e^{-2i\pi fTn_d} \mathfrak{F}\left\{x[n]\right\},$$

for integer-valued time shifts $n_d$. Now, by taking this fact as well as relation (3.1.3) into account, we have

$$\begin{aligned} W_N^R(f) = \mathfrak{F}\left\{w_N^R[n]\right\} &= \mathfrak{F}\left\{w_N^{cR}\left[n - \frac{N}{2}\right]\right\} \\ &= e^{-2i\pi f T \frac{N}{2}} \mathfrak{F}\left\{w_N^{cR}[n]\right\} \\ &= e^{-i\pi f T N} D_N(f). \end{aligned}$$

$\square$

## A.1.2 Indirect estimation methods

In this subsection we will prove Lemma 2. For this purpose, we first define the centered Bartlett window, which represents the easiest way of tapering the data symmetrically at both ends of the segment.

**Definition 27.** The *centered Bartlett (triangular) window* is defined by

$$w_N^{cB}[n] = \begin{cases} \frac{2|n|}{N} & 0 \leq |n| \leq \frac{N}{2} \\ 0 & \text{else}. \end{cases} \quad (A.1.1)$$

This definition at hand, we consider the proof of Lemma 2.

*Proof.* Properties of the correlogram with biased correlation.

1. Biased estimator.
   Relation (3.1.1) and the linearity of the expectation yield

$$\begin{aligned} \mathbb{E}\left\{\check{S}_x^C(f)\right\} &= T \sum_{m=-L}^{L} \mathbb{E}\{\check{r}_x[m]\} e^{-2i\pi f T m} \\ &= T \sum_{m=-L}^{L} \left(1 - \frac{|m|}{N}\right) r_x[m] e^{-2i\pi f T m} \\ &= T \sum_{m=-\infty}^{\infty} \left(1 - \frac{|m|}{N}\right) w_{2L}^{cR}[m] r_x[m] e^{-2i\pi f T m} \\ &= \mathfrak{F}\left\{r_x[m] w_{2L}^{cB}[m]\right\} \\ &= S_x(f) \star \frac{1}{L} D_{2L}^2\left(\frac{f}{2}\right), \end{aligned}$$

as the DTFT of the centered Bartlett window (A.1.1) is the *Fejer kernel* (compare Subsection 5.6 of Marple (1987))

$$W_N^{cB}(f) = \frac{2}{N} D_N^2\left(\frac{f}{2}\right).$$

Hereby, $D_N(f)$ denotes the Dirichlet kernel as defined in (3.1.4). We conclude that the PSD estimator is biased.

2. Asymptotically unbiased estimator.
   This follows immediately from step 1. For $N \to \infty$ the Dirichlet kernel approaches the Dirac delta[1], and thus

$$\lim_{N \to \infty} \mathbb{E}\left\{\check{S}_x^C(f)\right\} = \lim_{N \to \infty} S_x(f) \star \frac{1}{L} D_{2L}^2\left(\frac{f}{2}\right) = S_x(f).$$

$\square$

## A.1.3 Direct estimation methods

### A.1.3.1 Periodogram

Here we are concerned with the proof of the properties of the periodogram given in Lemma 3.

First we establish an alternative expression of the periodogram.

**Lemma 11** (Alternative representation of the periodogram). *The periodogram (3.1.8) equals the correlogram with biased correlation estimate (3.1.7), i.e.*

$$\hat{S}_x^P(f) = \check{S}_x^C(f). \qquad (A.1.2)$$

*Proof.* Alternative representation of the periodogram.

---

[1] For reasons of simplicity, we sloppily treat the Dirac delta as function and avoid the explicit consideration of convergence in the sense of distributions (generalized functions). We refer to Friedlander and Joshi (1999) for an introduction to the theory of distributions.

Compare Subsection 10.1 of Brockwell and Davis (1991) for the idea of the following proof. The periodogram is

$$\hat{S}_x^P(f) = \frac{T}{N} \left| \sum_{n=0}^{N-1} x[n] e^{-2i\pi f T n} \right|^2$$

$$= \frac{T}{N} \left( \sum_{n=0}^{N-1} x[n] e^{-2i\pi f T n} \right) \left( \sum_{n'=0}^{N-1} x[n']^* e^{+2i\pi f T n'} \right)$$

$$= \frac{T}{N} \sum_{n=0}^{N-1} \sum_{n'=0}^{N-1} x[n] x[n']^* e^{-2i\pi f T(n-n')}.$$

Substituting $m = n - n'$, we have to split up the obtained sum into a part of positive and into one of negative lag $m$.

For $m \geq 0$ we have

$$\hat{S}_x^P(f) = \frac{T}{N} \sum_{|m|<N-1} \sum_{n=m}^{N-1} x[n] x[n-m]^* e^{-2i\pi f T m}$$

$$= T \sum_{|m|<N-1} e^{-2i\pi f T m} \frac{1}{N} \sum_{n=0}^{N-m-1} x[n+m] x[n]^*$$

$$= T \sum_{|m|<N-1} e^{-2i\pi f T m} \check{r}_x[m],$$

where $\check{r}_x[m]$ the biased autocorrelation estimator.

For $m < 0$ we similarly obtain

$$\hat{S}_x^P(f) = \frac{T}{N} \sum_{|m|<N-1} \sum_{n=0}^{N+m-1} x[n] x[n-m]^* e^{-2i\pi f T m}$$

$$= T \sum_{|m|<N-1} e^{-2i\pi f T m} \frac{1}{N} \sum_{n=0}^{N+m-1} x[n] x[n-m]^*$$

$$= T \sum_{|m|<N-1} e^{-2i\pi f T m} \frac{1}{N} \sum_{n=0}^{N-|m|-1} x[n+|m|]^* x[n]^*$$

$$= T \sum_{|m|<N-1} e^{-2i\pi f T m} \check{r}_x[m].$$

These two cases taken together yield the desired identity

$$\hat{S}_x^P(f) = T \sum_{|m|<N-1} \check{r}_x[m] e^{-2i\pi f T m} = \check{S}_x^C(f),$$

which completes the proof. □

The alternative representation of the periodogram (A.1.2) allows us to prove its properties (Lemma 3):

*Proof.* Properties of the periodogram.
We have to prove 3 properties:

1. Biased estimator.
   According to Lemma 11, the periodogram equals the correlogram with biased estimator. Its expectation is thus, in complete analogy to the proof of Lemma 2,

$$\begin{aligned}
\mathbb{E}\{\hat{S}_x^P(f)\} &= \mathbb{E}\{\check{S}_x^C(f)\} \\
&= T \sum_{m=-(N-1)}^{N-1} \mathbb{E}\{\check{r}_x[m]\} e^{-2i\pi fmT} \\
&= T \sum_{m=-(N-1)}^{N-1} \left(1 - \frac{|m|}{N}\right) r_x[m] e^{-2i\pi fTm} \\
&= T \sum_{m=-\infty}^{\infty} \left(1 - \frac{|m|}{N}\right) w_{2N}^{cR}[m] r_x[m] e^{-2i\pi fTm} \\
&= \mathfrak{F}\{r_x[m] w_{2N}^{cB}[m]\} \\
&= S_x(f) \star \frac{1}{N} D_{2N}^2\left(\frac{f}{2}\right).
\end{aligned}$$

As in the proof of Lemma 2, $w_{2N}^{cR}[n]$ denotes the centered rectangular window and $w_{2N}^{cB}[n]$ the centered Bartlett window whose DTFT is the Fejer kernel. This implies that the periodogram is a biased estimator of the power spectral density.

2. Asymptotically unbiased estimator.
   This follows immediately from step 1 with the same reasoning as in the proof of Lemma 2: For $N \to \infty$ the Dirichlet kernel approaches

the Dirac delta[2], and thus

$$\lim_{N\to\infty} \mathbb{E}\{\hat{S}_x^P(f)\} = \lim_{N\to\infty} S_x(f) \star \frac{2}{2N-1} D_{2N-1}^2\left(\frac{f}{2}\right) = S_x(f).$$

3. Inconsistent estimator.

We proceed in two steps:

First, we first assume $x[n]$ to be (zero-mean) white Gaussian with variance $\mathbb{V}\{x[n]\} = \sigma^2$. Under this assumption the power spectral density of the white noise $x[n]$ takes the simple form

$$\begin{aligned}S_x(f) = \mathfrak{F}\{r_x[m]\} &= T\sum_{m=-\infty}^{\infty} r_x[m]\, e^{-2i\pi fTm}\\ &= T\sigma^2 + T\sum_{m\neq 0} r_x[m]\, e^{-2i\pi fTm}\\ &= T\sigma^2.\end{aligned}$$

Furthermore its fourth order statistical moment reduces to

$$\mathbb{E}\{x[k]\,x[l]\,x[m]\,x[n]\} = \begin{cases} \sigma^4 & (k=l) \text{ and } (m=n) \\ & \text{or } (k=m) \text{ and } (l=n) \\ & \text{or } (k=n) \text{ and } (l=m) \\ 0 & \text{else.} \end{cases}$$

Now let us investigate the properties of the periodogram under the assumption of Gaussian white noise. Its expectation then equals

$$\begin{aligned}\mathbb{E}\{\hat{S}_x^P(f)\} &= \frac{T}{N}\mathbb{E}\left\{\left(\sum_{n=0}^{N-1} x[n]\, e^{-2i\pi fTn}\right)\left(\sum_{n'=0}^{N-1} x[n']^*\, e^{+2i\pi fTn'}\right)\right\}\\ &= \frac{T}{N}\sum_{n=0}^{N-1} \mathbb{E}\{x[n]\,x[n]^*\}\\ &= T\sigma^2.\end{aligned}$$

---

[2] Again, we do not explicitly indicate convergence in the sense of distributions.

Using the fourth order statistical moment derived above, we can express its autocorrelation as $\mathbb{E}\left\{\hat{S}_x^P(f_1)\hat{S}_x^P(f_2)\right\} = \ldots$

$$= \mathbb{E}\left\{\frac{1}{NT}[X(f_1)X(f_1)^*]\frac{1}{NT}[X(f_2)X(f_2)^*]\right\}$$

$$= \frac{T^2}{N^2}\sum_{k=0}^{N-1}\sum_{l=0}^{N-1}\sum_{m=0}^{N-1}\sum_{n=0}^{N-1}\mathbb{E}\{x[k]x[l]x[m]x[n]\}\cdot e^{2i\pi T[f_1(k-l)+f_2(m-n)]}$$

$$= T^2\sigma^4\left[1+\left(\frac{\sin(\pi NT(f_1+f_2))}{N\sin(\pi T(f_1+f_2))}\right)^2 + \left(\frac{\sin(\pi NT(f_1-f_2))}{N\sin(\pi T(f_1-f_2))}\right)^2\right].$$

Collecting all these results, the variance of the periodogram is

$$\mathbb{V}\{\hat{S}_x^P(f)\} = \mathbb{E}\{\hat{P}_x^P(f)\hat{P}_x^P(f)\} - \mathbb{E}\{\hat{P}_x^P(f)\}\mathbb{E}\{\hat{P}_x^P(f)\}$$

$$= T^2\sigma^4\left[1+\left(\frac{\sin(\pi NT 2f)}{N\sin(\pi T 2f)}\right)^2+1\right] - T\sigma^2\cdot T\sigma^2$$

$$= S_x^2(f)\left[1+\left(\frac{\sin(2\pi fTN)}{N\sin(2\pi fT)}\right)^2\right].$$

Clearly, this expression does not converge to zero with increasing $N$, but remains in the order of the squared spectral power density. This implies that in case of a white Gaussian signal the periodogram is inconsistent.

In case of non-white signals a similar argumentation is possible, see Jenkins and Watts (1968) for technical details. This completes the proof that the periodogram is an inconsistent estimator of the power spectral density.

□

### A.1.3.2 Welch method

Here, we are concerned with the proof of the properties of the Welch periodogram (Lemma 4).

Again, first we establish an alternative expression of the Welch periodogram which we will then need. Note that the following proof is in complete analogy to the one of the relation of the periodogram to the correlogram with biased correlation estimator (Lemma 11), but slightly more technically demanding.

**Lemma 12** (Alternative representation of the Welch periodogram). *The Welch periodogram (3.1.10) can be expressed as*

$$\hat{S}_x^W(f) = \frac{1}{U} \cdot \frac{1}{P} \sum_{p=1}^{P-1} \check{S}_{wx}^{C(p)}(f), \quad (A.1.3)$$

*where $\check{S}_{wx}^{C(p)}(f)$ denotes the correlogram with biased estimate of the windowed data samples belonging to the pth segment $x^{(p)}[n] = w[n]x[pS+n]$.*

*Proof.* Alternative representation of the Welch periodogram.

First, we expand the Welch periodogram as

$$\begin{aligned}
\hat{S}_x^W(f) &= \frac{1}{P} \sum_{p=0}^{P-1} \tilde{S}_x^{(p)}(f) \\
&= \frac{T}{UP} \sum_{p=0}^{P-1} \frac{1}{D} \left| \sum_{n=0}^{D-1} x^{(p)}[n] e^{-2i\pi f T n} \right|^2 \\
&= \frac{T}{UP} \sum_{p=0}^{P-1} \frac{1}{D} \left( \sum_{n=0}^{D-1} w[n] x[pS+n] e^{-2i\pi f T n} \right) \\
&\quad \cdot \left( \sum_{n'=0}^{D-1} w[n'] x[pS+n']^* e^{+2i\pi f T n'} \right) \\
&= \frac{T}{UP} \sum_{p=0}^{P-1} \frac{1}{D} \sum_{n=0}^{D-1} \sum_{n'=0}^{D-1} w[n] w[n']^* x[pS+n] x[pS+n']^* e^{-2i\pi f T(n-n')}.
\end{aligned}$$

Second, we substitute $m = n - n'$ and split up the obtained sum into a part of positive and into one of negative lag $m$:

For $m \geqslant 0$ we have $\hat{S}_x^W(f) = \ldots$

$$= \frac{T}{UP} \cdot \frac{1}{D} \sum_{|m|<D-1} \sum_{n=m}^{D-1} w[n]\, x[pS+n]\, x[pS+n-m]^* w[n-m]^* e^{-2i\pi fTm}$$

$$= \frac{T}{UP} \sum_{|m|<D-1} e^{-2i\pi fTm} \frac{1}{D} \sum_{n=0}^{D-m-1} w[n+m]\, x[pS+n+m]\, x[pS+n]^* w[n]^*$$

$$= \frac{T}{UP} \sum_{|m|<D-1} e^{-2i\pi fTm} \,\check{r}_{wx}^{(p)}[m],$$

where $\check{r}_{wx}^{(p)}[m]$ denotes the biased autocorrelation estimator based on samples of the pth segment $x^{(p)}[n] = w[n]\, x[pS+n]$.

For $m < 0$ we similarly obtain $\hat{S}_x^W(f) = \ldots$

$$= \frac{T}{UPD} \sum_{|m|<D-1} \sum_{n=0}^{D+m-1} w[n]\, x[pS+n]\, x[pS+n-m]^* w[n-m]^* e^{-2i\pi fTm}$$

$$= \frac{T}{UP} \sum_{|m|<D-1} e^{-2i\pi fTm} \frac{1}{D} \sum_{n=0}^{D+m-1} w[n]\, x[pS+n]\, x[pS+n-m]^* w[n-m]^*$$

$$= \frac{T}{UP} \sum_{|m|<D-1} e^{-2i\pi fTm} \,\check{r}_{wx}^{(p)}[m].$$

This leads to the identical sum of autocorrelation estimators in both cases, which now involves $\check{r}_{wx}^{(p)}[m]$, i.e. the biased autocorrelation estimator based on samples of the pth segment $x^{(p)}[n] = w[n]\, x[pS+n]$. Thus, we obtain the desired identity

$$\hat{S}_x^W(f) = \frac{1}{U} \cdot \frac{1}{P} \sum_{p=0}^{P-1} T \sum_{|m|<D-1} \check{r}_{wx}^{(p)}[m]\, e^{-2i\pi fTm}$$

$$= \frac{1}{U} \cdot \frac{1}{P} \sum_{p=0}^{P-1} \check{S}_{wx}^{C(p)}(f).$$

$\square$

The alternative representation of the Welch periodogram (A.1.3) allows us to prove its properties (Lemma 4).

*Proof.* Properties of the Welch periodogram.
We have to prove 3 properties:

1. Biased estimator.

   First we derive the expectation of the biased autocorrelation of the windowed data $\check{r}_{wx}^{(p)}[m]$. Exploiting the linearity of the expectation and the fact that the window $w[n]$ is a deterministic signal, we can express it as

   $$\begin{aligned}
   \mathbb{E}\left\{\check{r}_{wx}^{(p)}[m]\right\} &= \mathbb{E}\left\{\frac{1}{D}\sum_{n=0}^{D-m-1} w[n+m]\,x[pS+n+m]\,x[pS+n]\,w[n]\right\} \\
   &= \frac{1}{D}\sum_{n=0}^{D-m-1} w[n+m]\,\mathbb{E}\{x[pS+n+m]\,x[pS+n]\}\,w[n] \\
   &= r_x^{(p)}[m]\,\frac{1}{D}\sum_{n=0}^{D-m-1} w[n+m]\,w[n] \\
   &= r_x^{(p)}[m]\,\check{r}_w[m].
   \end{aligned} \qquad (A.1.4)$$

   The expectation of the Welch periodogram (3.1.8) is thus, according to (A.1.3) and (A.1.4),

   $$\begin{aligned}
   \mathbb{E}\{\hat{S}_x^W(f)\} &= \mathbb{E}\left\{\frac{1}{U}\cdot\frac{1}{P}\sum_{p=1}^{P-1} \check{P}_{wx}^{C(p)}(f)\right\} \\
   &= \frac{T}{U}\frac{1}{P}\sum_{p=0}^{P-1}\sum_{m=-(D-1)}^{D-1} \mathbb{E}\{\check{r}_{wx}^{(p)}[m]\}\,e^{-2i\pi fmT} \\
   &= \frac{T}{U}\frac{1}{P}\sum_{p=0}^{P-1}\sum_{m=-(D-1)}^{D-1} r_x^{(p)}[m]\,\check{r}_w[m]\,e^{-2i\pi fTm}.
   \end{aligned}$$

   As the window $w[n]$ is a signal of finite length $2D-1$, its autocorrelation estimate $\check{r}_w[m]$ only has $2D-1$ non-zero lags, thus $\check{r}_w[m] = 0$ for $m > D-1$.

This allows us to extend the above sum to

$$\begin{aligned}
\mathbb{E}\{\hat{S}_x^W(f)\} &= \frac{T}{U}\frac{1}{P}\sum_{p=0}^{P-1}\sum_{m=-\infty}^{\infty} r_x^{(p)}[m]\,\check{r}_w[m]\,e^{-2i\pi f T m} \\
&= \frac{1}{U}\frac{1}{P}\sum_{p=0}^{P-1} \mathfrak{F}\left\{r_x^{(p)}[m]\,\check{r}_w[m]\right\} \\
&= \frac{1}{U}\frac{1}{P}\sum_{p=0}^{P-1} S_x^{(p)}(f) \star \check{S}_w^C(f),
\end{aligned}$$

where, according to definition (3.1.7), $\check{S}_w^C(f)$ denotes the correlogram with biased correlation estimate of the window $w[n]$.

Exploiting the distributive law for the convolution, we have

$$\mathbb{E}\{\hat{S}_x^W(f)\} = \frac{1}{U}\underbrace{\left(\frac{1}{P}\sum_{p=0}^{P-1} S_x^{(p)}(f)\right)}_{S_x(f)} \star \check{S}_w^C(f).$$

As the correlogram with biased correlation estimate equals the periodogram according to (A.1.2), we finally obtain

$$\begin{aligned}
\mathbb{E}\{\hat{S}_x^W(f)\} &= \frac{1}{U} S_x(f) \star \hat{S}_w^P(f) \\
&= S_x(f) \star \frac{T}{DU}|W(f)|^2,
\end{aligned}$$

where $W(f) = \mathfrak{F}\{w[n]\}$ is the Fourier transform of the window $w[n]$. This implies that the Welch periodogram is a biased estimator of the power spectral density.

2. Asymptotically unbiased estimator.

First note that the definition of U ensures the correct scaling: We have

$$U = \frac{T}{D}\sum_{n=0}^{D-1} w[n]^2 = \int_{-\frac{1}{2T}}^{\frac{1}{2T}} \frac{T}{D}|W(f)|^2\,df,$$

which implies

$$\frac{T}{DU}\int_{-\frac{1}{2T}}^{\frac{1}{2T}} |W(f)|^2 = 1.$$

Therefore, for $N \to \infty$, the scaled impulse response of the window approaches the Dirac delta[3], and thus

$$\lim_{N\to\infty} \mathbb{E}\left\{\hat{S}_x^W(f)\right\} = \lim_{N\to\infty} S_x(f) \star \frac{T}{DU}|W(f)|^2 = S_x(f).$$

3. Inversely proportional relation of estimator variance and segment number.

As this proof is of technical nature, we only outline its idea (Welch 1967). Assuming segment independence (although a non-zero overlap will lead to dependence between adjacent segments), it can be shown that

$$\mathbb{V}\left\{\hat{S}_x^W(f)\right\} \propto \frac{S_x^2(f)}{P}.$$

We refer to Welch (1967) for further details.

$\square$

## A.2 Granger causality

We refer to Eichler (2007) for a proof of the characterization of conditional Granger causality via AR coefficients.

## A.3 Dependency measures

### A.3.1 Partial Coherence

We prove formula (3.4.5) in two steps, whereupon the first is formulated as a lemma. It claims a well-known relation, compare e.g. Brillinger (2001).

**Lemma 13.** *The partial cross-spectral density (3.4.3) can be obtained by calculating*

$$S_{i,j|\mathbf{y}}(f) = S_{ij}(f) - \mathbf{S}_{i\mathbf{y}}(f)\mathbf{S}_{\mathbf{yy}}^{-1}(f)\mathbf{S}_{\mathbf{y}j}(f), \qquad (\text{A.3.1})$$

*where $S_{ij}(f)$ is the (i,j)-entry of the spectral matrix $\mathbf{S}(f)$, and the terms $\mathbf{S}_{i\mathbf{y}}(f)$, $\mathbf{S}_{\mathbf{yy}}(f)$ and $\mathbf{S}_{\mathbf{y}j}(f)$ are the respective block matrices of $\mathbf{S}(f)$.*

---

[3] Again, we do not explicitly indicate convergence in the sense of distributions.

*Proof.* Partialization of cross-spectral density.

We consider the partial cross-spectral density $S_{i,j|\mathbf{y}}(f)$ given in (3.4.3). According to (3.4.2) we have

$$\begin{aligned} S_{i,j|\mathbf{y}}(f) &= \mathfrak{F}\mathbb{E}\{\varepsilon_i[n+s]\varepsilon_j^T[n]\} \\ &= \mathfrak{F}\mathbb{E}\{(x_i[n+s]-\mathbf{d}_i^T*\mathbf{y}[n+s])(x_j[n]-\mathbf{d}_j^T*\mathbf{y}[n])^T\} \\ &= \mathfrak{F}\mathbb{E}\{x_i[n+s]x_j[n]\} - \mathfrak{F}\mathbb{E}\{x_i[n+s](\mathbf{y}^T[n]*\mathbf{d}_j)\} - \\ &\quad -\mathfrak{F}\mathbb{E}\{(\mathbf{d}_i*\mathbf{y}[n+s])x_j[n]\} \\ &\quad +\mathfrak{F}\mathbb{E}\{(\mathbf{d}_i^T*\mathbf{y}[n+s])(\mathbf{y}^T[n]*\mathbf{d}_j)\}. \end{aligned}$$

With $\mathbf{D}_i(f) = \mathfrak{F}\mathbf{d}_i[\tau]$ and $\mathbf{D}_j(f) = \mathfrak{F}\mathbf{d}_j[\tau]$ we have

$$\begin{aligned} S_{i,j|\mathbf{y}}(f) &= S_{ij}(f) - \mathbf{S}_{i\mathbf{y}}(f)\mathbf{D}_j(f) \\ &\quad - \mathbf{D}_i^T(f)\mathbf{S}_{\mathbf{y}j}(f) + \mathbf{D}_i^T(f)\mathbf{S}_{\mathbf{yy}}(f)\mathbf{D}_j(f). \end{aligned} \quad (A.3.2)$$

According to the Wiener filter formula, the filters can be expressed as

$$\begin{cases} \mathbf{D}_i^T(f) = \mathbf{S}_{i\mathbf{y}}(f)\mathbf{S}_{\mathbf{yy}}^{-1}(f) \\ \mathbf{D}_j^T(f) = \mathbf{S}_{\mathbf{y}j}^T(f)\mathbf{S}_{\mathbf{yy}}^{-1}(f) \end{cases}.$$

Plugging this representation into (A.3.2) gives

$$\begin{aligned} S_{i,j|\mathbf{y}}(f) &= S_{ij}(f) - \mathbf{S}_{i\mathbf{y}}(f)\left(\mathbf{S}_{\mathbf{y}j}(f)^T\mathbf{S}_{\mathbf{yy}}^{-1}(f)\right)^T \\ &\quad -\mathbf{S}_{i\mathbf{y}}(f)\mathbf{S}_{\mathbf{yy}}^{-1}(f)\mathbf{S}_{\mathbf{y}j}(f) + \mathbf{S}_{i\mathbf{y}}(f)\mathbf{S}_{\mathbf{yy}}^{-1}(f)\mathbf{S}_{\mathbf{yy}}(f)\mathbf{S}_{\mathbf{yy}}^{-1}(f)\mathbf{S}_{\mathbf{y}j}(f) \\ &= S_{ij}(f) - \mathbf{S}_{i\mathbf{y}}(f)\mathbf{S}_{\mathbf{yy}}^{-1}(f)\mathbf{S}_{\mathbf{y}j}(f), \end{aligned}$$

which is the desired result. □

This allows us to prove the computational formula of the partial coherence (3.4.5).

*Proof.* Computational formula of PC.

First note that, due to relation (A.3.1), we only have to prove that the numerator of the computational formula (3.4.5) gives the right-hand-side of equation (A.3.1).

Therefore, let $i = 1$ and $j = 2$ without loss of generality. In order to calculate $M_{1,2}(f)$, we remove row 1 and column 2 of $\mathbf{S}(f)$ and obtain

$$\begin{pmatrix} S_{21}(f) & S_{23}(f) & S_{24}(f) & \cdots \\ S_{31}(f) & S_{33}(f) & S_{34}(f) & \cdots \\ S_{41}(f) & S_{43}(f) & S_{44}(f) & \cdots \\ \vdots & \vdots & \vdots & \ddots \end{pmatrix} = \begin{pmatrix} S_{21}(f) & \mathbf{S_{2y}}(f) \\ \mathbf{S_{y1}}(f) & \mathbf{S_{yy}}(f) \end{pmatrix}$$

Its determinant is, according to linear algebra,

$$\det \begin{pmatrix} S_{21}(f) & \mathbf{S_{2y}}(f) \\ \mathbf{S_{y1}}(f) & \mathbf{S_{yy}}(f) \end{pmatrix} = \left| S_{21}(f) - \mathbf{S_{2y}}(f)\mathbf{S_{yy}}^{-1}(f)\mathbf{S_{y1}}(f) \right| \cdot \left| \mathbf{S_{yy}}(f) \right|.$$

The factor $|\mathbf{S_{yy}}(f)^2|$ is present in both the numerator and the denominator of (3.4.5) and thus cancels out, which completes the proof. □

### A.3.2 Directed Transfer Function

Here we prove the relation of DTF and bivariate Granger causality as stated in Lemma 6.

*Proof.* Link to Granger causality in the bivariate case.

Let $\mathbf{x}[n] = (x_1[n], x_2[n])^T$ a stationary bivariate signal modeled within the autoregressive framework (3.2.1). Thus, the matrices $\mathbf{A}(f)$ and $\mathbf{H}(f) \in \mathbb{C}^{2 \times 2}$.

In order to test on zero, it is sufficient to consider the numerator of DTF (3.4.6), $|H_{ij}(f)|^2$. The latter is, according to lemma of matrix inversion,

$$|H_{ij}(f)|^2 = \left| [A(f)]_{ij} \right|^2 = \frac{|A_{ij}(f)|^2}{|\mathbf{A}(f)|^2}, \; i \neq j.$$

Thus, $\gamma_{ij}^2(f) = 0 \; \forall f$ is equivalent to $A_{ij}(f) = 0 \; \forall f$, which completes the proof. □

Note that this proof can be generalized, if $x[n]$ of dimension $K$ consists of two component signals $x_I[n]$ (dimension M) and $x_J[n]$ (dimension N) with $M + N = K$, as the matrix inversion lemma also applies to block-wise inversion.

## A.3.3 Partial Directed Coherence

We start with the construction of PDC.

First, consider the proof of Lemma 7.

*Proof.* Partial Coherence Function and PC.

According to Dahlhaus (2000), the partialized spectrum can be obtained by considering the re-normalized inverse spectrum. The inverse spectrum is

$$\begin{aligned} \mathbf{S}^{-1}(f) &= \left(\mathbf{H}\Sigma\mathbf{H}^H\right)^{-1} \\ &= \mathbf{H}^{-H}\Sigma^{-1}\mathbf{H}^{-1} \\ &= \mathbf{A}^H\Sigma^{-1}\mathbf{A}, \end{aligned}$$

and the (i,j)-element of this expression is

$$\left[\mathbf{S}^{-1}(f)\right]_{ij} = \mathbf{a}_{\cdot i}^H(f)\Sigma^{-1}\mathbf{a}_{\cdot j}(f). \tag{A.3.3}$$

The Partial Coherence (3.4.4) is thus given by

$$R_{i,j|\mathbf{y}}^2(f) = \frac{\mathbf{a}_{\cdot i}^H(f)\Sigma^{-1}\mathbf{a}_{\cdot j}(f)}{\sqrt{\mathbf{a}_{\cdot i}^H(f)\Sigma^{-1}\mathbf{a}_{\cdot i}(f)}\sqrt{\mathbf{a}_{\cdot j}^H(f)\Sigma^{-1}\mathbf{a}_{\cdot j}(f)}} = \kappa_{i,j}(f), \tag{A.3.4}$$

as the re-normalization factor of (A.3.3) cancels out in the numerator and the denominator. □

Second, consider the proof of Lemma 8

*Proof.* Factorization of the Partial Coherence Function.

First, for simplicity reasons, let us consider a special case. For a bivariate system, we have

$$\Pi = \begin{pmatrix} \pi_{\cdot 1} & \pi_{\cdot 2} \end{pmatrix} = \begin{pmatrix} \dfrac{A_{11}(f)}{\sqrt{\mathbf{a}_{\cdot 1}^H(f)\Sigma^{-1}\mathbf{a}_{\cdot 1}(f)}} & \dfrac{A_{12}(f)}{\sqrt{\mathbf{a}_{\cdot 2}^H(f)\Sigma^{-1}\mathbf{a}_{\cdot 2}(f)}} \\ \dfrac{A_{21}(f)}{\sqrt{\mathbf{a}_{\cdot 1}^H(f)\Sigma^{-1}\mathbf{a}_{\cdot 1}(f)}} & \dfrac{A_{22}(f)}{\sqrt{\mathbf{a}_{\cdot 2}^H(f)\Sigma^{-1}\mathbf{a}_{\cdot 2}(f)}} \end{pmatrix}.$$

Let $i = 1$, $j = 2$ without loss of generality. Obviously,

$$\pi_{\cdot 1}^H(f)\Sigma^{-1}\pi_{\cdot 2}(f) =$$

$$= \begin{pmatrix} \dfrac{A_{11}(f)}{\sqrt{\mathbf{a}_{\cdot 1}^H(f)\Sigma^{-1}\mathbf{a}_{\cdot 1}(f)}} \\ \dfrac{A_{21}(f)}{\sqrt{\mathbf{a}_{\cdot 1}^H(f)\Sigma^{-1}\mathbf{a}_{\cdot 1}(f)}} \end{pmatrix}^H \Sigma^{-1} \begin{pmatrix} \dfrac{A_{12}(f)}{\sqrt{\mathbf{a}_{\cdot 2}^H(f)\Sigma^{-1}\mathbf{a}_{\cdot 2}(f)}} \\ \dfrac{A_{22}(f)}{\sqrt{\mathbf{a}_{\cdot 2}^H(f)\Sigma^{-1}\mathbf{a}_{\cdot 2}(f)}} \end{pmatrix}$$

$$= \dfrac{1}{\sqrt{\mathbf{a}_{\cdot 1}^H(f)\Sigma^{-1}\mathbf{a}_{\cdot 1}(f)}} \begin{pmatrix} A_{11}^* & A_{21}^* \end{pmatrix} \Sigma^{-1} \begin{pmatrix} A_{12} \\ A_{22} \end{pmatrix} \dfrac{1}{\sqrt{\mathbf{a}_{\cdot 2}^H(f)\Sigma^{-1}\mathbf{a}_{\cdot 2}(f)}}$$

$$= \dfrac{\mathbf{a}_{\cdot 1}^H(f)\Sigma^{-1}\mathbf{a}_{\cdot 2}(f)}{\sqrt{\mathbf{a}_{\cdot 1}^H(f)\Sigma^{-1}\mathbf{a}_{\cdot 1}(f)}\sqrt{\mathbf{a}_{\cdot 2}^H(f)\Sigma^{-1}\mathbf{a}_{\cdot 2}(f)}}$$

$$= \kappa_{1,2}(f).$$

The proof for the general case $K > 2$ is in complete analogy. $\square$

An important property of PDC is its ability to indicate conditional Granger causality, i.e. causality in the multivariate case (unlike DTF).

*Proof.* Link to Conditional Granger Causality.

Immediate, according to Definition 7, also compare Subsection 3.3.2.
$\square$

Finally, we prove that PDC is well-defined under reasonable conditions.

*Proof.* Well-definedness of PDC.

Under the stability condition (3.2.2), the eigenvalues $\lambda_i$ of the polynomial matrix $\mathbf{A}(z) = \mathbf{1}_{K \times K} - \mathbf{A}[1]z - \ldots - \mathbf{A}[p]z^p$, $z \in \mathbb{C}$, are all different from zero. Consequently, $\mathbf{A}(f)^H \mathbf{A}(f) > 0$.

Let $\mathbf{e}_j \triangleq \begin{pmatrix} 0 & \ldots & 0 & 1 & 0 & \ldots & 0 \end{pmatrix}^T$ the jth base vector with a single 1 on position $j$ and $\|\cdot\|$ the Euclidean norm.

With this notation, the denominator of the PDC is

$$\sum_{n=1}^{K} |A_{nj}(f)|^2 = \|\mathbf{A}(f)\mathbf{e}_j\|^2$$
$$= (\mathbf{A}(f)\mathbf{e}_j)^H (\mathbf{A}(f)\mathbf{e}_j)$$
$$= \mathbf{e}_j \underbrace{\mathbf{A}(f)^H \mathbf{A}(f)}_{>0} \mathbf{e}_j > 0.$$

The strictly positive denominator thus assures the well-definedness of PDC. □

## A.4 Signal detection

First, we prove Theorem 2.

*Proof.* Invariance of decision rule. We proof both directions:

- First, let us proof that $\phi(x) = \phi_I(T(x))$ is sufficient for $\phi$ to be invariant, i.e. $\phi(g(x)) = \phi(x)$.
  Assume that $\phi(g(x)) = \phi_I(T(x))$. Then

  $$\phi(g(x)) = \phi_I(g(T(x)))$$
  $$= \phi_I(T(x)) \ (T(x) \text{ is maximal invariant, property 1})$$
  $$= \phi(x).$$

  Thus, $\phi(g(x)) = \phi(x)$, which is the desired result.

- Alternatively, let us prove the equivalent condition that the implication $T(x_1) = T(x_2) \Rightarrow \phi(x_1) = \phi(x_2)$ is sufficient for $\phi$ to be invariant. Assume that the implication $T(x_1) = T(x_2) \Rightarrow \phi(x_1) = \phi(x_2)$ holds. As $T(x)$ is maximal invariant, we let $x_2 = g(x_1)$. Then the assumption yields $\phi(x_1) = \phi(g(x_1))$, which is the desired result.

- Finally, we prove the opposite implication, i.e. that the implication $T(x_1) = T(x_2) \Rightarrow \phi(x_1) = \phi(x_2)$ is necessary for $\phi$ to be invariant.

Assume that $\phi(g(x)) = \phi(x)$. As $T(x)$ is maximal invariant (property 2), we have the implication $T(x_1) = T(x_2) \Rightarrow x_2 = g(x_1)$. Therefrom, and by the assumption, we have

$$\phi(x_2) = \phi(g(x_1)) = \phi(x_1).$$

Thus, the implication $T(x_1) = T(x_2) \Rightarrow \phi(x_1) = \phi(x_2)$ holds, which is the desired result.

$\square$

Finally, we detail the derivation of the matched subspace filter (3.5.3). For this purpose we start with the proof that the hypothesis testing problem is invariant.

*Proof.* Problem invariance with respect to $\mathscr{G} = \{g : g(\mathbf{x}) = \mathbf{Q}_\mathscr{S}\mathbf{x} + \mathbf{z}, \mathbf{z} \in S^\perp\}$.

- We have to show the invariance of the transformation group, i.e. that $p_1(\mathbf{x}_1, \theta) = p(\mathbf{x}_1, \theta_1)$ for $\mathbf{x}_1 = g(\mathbf{x})$.
First note that, as $\mathbf{Q}_\mathscr{S}\mathbf{H}\theta \in \mathscr{S}$,

$$\begin{aligned}\mathbf{Q}_\mathscr{S}\mathbf{H}\theta &= \mathbf{P}_\mathscr{S}\mathbf{Q}_\mathscr{S}\mathbf{H}\theta \\ &= \mathbf{H}(\mathbf{H}^T\mathbf{H})^{-1}\mathbf{H}^T\mathbf{Q}_\mathscr{S}\mathbf{H}\theta \\ &= \mathbf{H}\mathbf{H}^\#\mathbf{Q}_\mathscr{S}\mathbf{H}\theta \\ &= \mathbf{H}\theta_1,\end{aligned}$$

where $\mathbf{H}^\#$ denotes the Moore-Penrose pseudo-inverse of $\mathbf{H}$, and $\theta_1 = \mathbf{H}^\#\mathbf{Q}_s\mathbf{H}\theta$.
Therefore, we have

$$\begin{aligned}g(\mathbf{x}) = \mathbf{Q}_\mathscr{S}\mathbf{x} + \mathbf{z} &= \mathbf{Q}_\mathscr{S}\mathbf{s} + \mathbf{Q}_\mathscr{S}\mathbf{v} + \mathbf{Q}_\mathscr{S}\mathbf{w} + \mathbf{z} \\ &= \mathbf{Q}_\mathscr{S}\mathbf{H}\theta + \mathbf{v} + \mathbf{z} + \mathbf{Q}_\mathscr{S}\mathbf{w} \\ &= \mathbf{H}\theta_1 + \mathbf{v}_1 + \mathbf{Q}_\mathscr{S}\mathbf{w}\end{aligned}$$

with $\mathbf{v}_1 = \mathbf{v} + \mathbf{z}$. As $\mathbf{Q}_{\mathscr{S}}\mathbf{w} \sim \mathscr{N}(0, \sigma^2 \mathbf{I})$ due to the orthogonality of $\mathbf{Q}_{\mathscr{S}}$, we have $g(\mathbf{x}) \sim \mathscr{N}(\mathbf{H}\boldsymbol{\theta}_1 + \mathbf{v}_1, \sigma^2 \mathbf{I})$.

Thus, recalling that $\mathbf{x} \sim \mathscr{N}(\mathbf{H}\boldsymbol{\theta}_1 + \mathbf{v}_1, \sigma^2 \mathbf{I})$, we have shown that $p_1(\mathbf{x}_1, \boldsymbol{\theta}) = p(\mathbf{x}_1, \boldsymbol{\theta})$ with the induced transformation $\bar{g}$

$$(\boldsymbol{\theta}_1, \mathbf{v}_1) = \bar{g}(\boldsymbol{\theta}, \mathbf{v}) = (\mathbf{H}^{\#}\mathbf{Q}_{\mathscr{S}}\mathbf{H}\boldsymbol{\theta}, \mathbf{v} + \mathbf{z}).$$

- In addition we have to show that $\bar{g}$ preserves the dichotomy $\Theta = \Theta_0 \cup \Theta_1$, i.e. that the two conditions hold:

$$\boldsymbol{\theta} \in \Theta_0 \Leftrightarrow \bar{g}(\boldsymbol{\theta}) \in \Theta_0,$$
$$\boldsymbol{\theta} \in \Theta_1 \Leftrightarrow \bar{g}(\boldsymbol{\theta}) \in \Theta_1.$$

We have

$$\|\boldsymbol{\theta}\|^2 = 0 \Leftrightarrow \|\boldsymbol{\theta}_1\|^2 = 0,$$
$$\|\boldsymbol{\theta}\|^2 > 0 \Leftrightarrow \|\boldsymbol{\theta}_1\|^2 > 0,$$

which completes the proof that the hypothesis testing problem is invariant with respect to $g(\mathbf{x})$.

$\square$

Second, we prove that the considered statistic of the matched subspace filter (3.5.3) is maximal invariant.

*Proof.* Maximality of the invariance statistic $T(\mathbf{x}) = \|\mathbf{P}_{\mathscr{S}}\mathbf{x}\|^2$.

- First, we have to prove that $T(g(\mathbf{x})) = T(\mathbf{x})$ for all $g \in \mathscr{G}$ (property 1). Thus, we develop, by taking into account the properties of orthogonal projections,

$$\begin{aligned} T(g(\mathbf{x})) &= \|\mathbf{P}_{\mathscr{S}}(\mathbf{Q}_{\mathscr{S}}\mathbf{x} + \mathbf{z})\|^2 = \|\mathbf{P}_{\mathscr{S}}\mathbf{U}\mathbf{Q}\mathbf{U}^T\mathbf{x} + \mathbf{P}_{\mathscr{S}}\mathbf{P}_{\mathscr{S}^{\perp}}\mathbf{x}\|^2 \\ &= \mathbf{x}^T\mathbf{U}\mathbf{Q}^T\mathbf{U}^T\mathbf{P}_{\mathscr{S}}^T\mathbf{P}_{\mathscr{S}}\mathbf{U}\mathbf{Q}\mathbf{U}^T\mathbf{x} \\ &= \mathbf{x}^T\mathbf{U}\mathbf{Q}^T\mathbf{Q}\mathbf{U}^T\mathbf{x} = \mathbf{x}^T\mathbf{U}\mathbf{U}^T\mathbf{x} = \mathbf{x}^T\mathbf{P}_{\mathscr{S}}\mathbf{x} \\ &= \mathbf{x}^T\mathbf{P}_{\mathscr{S}}^2\mathbf{x} = \mathbf{P}_{\mathscr{S}}\mathbf{x} = T(\mathbf{x}), \end{aligned}$$

which is the desired result.

– Next, we have to prove that the implication $T(\mathbf{x}_1) = T(\mathbf{x}_2) \Rightarrow \mathbf{x}_2 = g(\mathbf{x}_1)$ holds for some $g \in \mathscr{G}$ (property 2).
For $i = \{1,2\}$, let $\mathbf{x}_i = \mathbf{x}_i^{\mathscr{S}} + \mathbf{x}_i^{\mathscr{S}^\perp}$, where $\mathbf{x}_i^{\mathscr{S}} = \mathbf{P}_{\mathscr{S}}\mathbf{x}_i$ and $\mathbf{x}_i^{\mathscr{S}^\perp} = \mathbf{P}_{\mathscr{S}^\perp}\mathbf{x}_i$. With this notation we have, by taking into account the properties of orthogonal projections,

$$
\begin{aligned}
T(\mathbf{x}_1) = T(\mathbf{x}_2) &\Rightarrow \|\mathbf{P}_{\mathscr{S}}\mathbf{x}_1\|^2 = \|\mathbf{P}_{\mathscr{S}}\mathbf{x}_2\|^2 \\
&\Rightarrow \left\|\mathbf{P}_{\mathscr{S}}\mathbf{x}_1^{\mathscr{S}} + \mathbf{P}_{\mathscr{S}}\mathbf{x}_1^{\mathscr{S}^\perp}\right\|^2 = \left\|\mathbf{P}_{\mathscr{S}}\mathbf{x}_2^{\mathscr{S}} + \mathbf{P}_{\mathscr{S}}\mathbf{x}_2^{\mathscr{S}^\perp}\right\|^2 \\
&\Rightarrow \left\|\mathbf{x}_1^{\mathscr{S}}\right\|^2 = \left\|\mathbf{x}_2^{\mathscr{S}}\right\|^2 \\
&\Rightarrow \mathbf{x}_2^{\mathscr{S}} = \mathbf{Q}_{\mathscr{S}}\mathbf{x}_1^{\mathscr{S}} \\
&\Rightarrow \mathbf{x}_2 = \mathbf{Q}_{\mathscr{S}}\mathbf{x}_1^{\mathscr{S}} + \mathbf{x}_2^{\mathscr{S}^\perp} = \mathbf{Q}_{\mathscr{S}}\mathbf{x}_1 - \mathbf{Q}_S\mathbf{x}_1^{\mathscr{S}^\perp} + \mathbf{x}_2^{\mathscr{S}^\perp} \\
&\Rightarrow \mathbf{x}_2 = \mathbf{Q}_{\mathscr{S}}\mathbf{x}_1 - \mathbf{x}_1^{\mathscr{S}^\perp} + \mathbf{x}_2^{\mathscr{S}^\perp} = \mathbf{Q}_{\mathscr{S}}\mathbf{x}_1 + \mathbf{z},
\end{aligned}
$$

with $\mathbf{z} \in S^\perp$. Thus, the initial assumption implies $\mathbf{x}_2 = g(\mathbf{x}_1)$, which is the desired result.

□

# Bibliography

Abetz, L., Jacoby, A., Baker, G. and McNulty, P.: 2000, Patient-Based Assessments of Quality of Life in Newly Diagnosed Epilepsy Patients: Validation of the NEWQOL, *Epilepsia* **41**(9), 1119–1128.

Adak, S.: 1998, Time-dependent Spectral Analysis of Nonstationary Time Series, *Journal of the American Statistical Association* **93**(444), 1488–1501.

Agarwal, R. and Gotman, J.: 1999, Adaptive segmentation of electroencephalographic data using a nonlinear energy operator, *Proceedings of the IEEE ISCAS* pp. IV 199–202.

Akiyama, T., McCoy, B., Go, C., Ochi, A., Elliott, I., Akiyama, M., Donner, E., Weiss, S., Snead III, O. C., Rutka, J., Drake, J. and Otsubo, H.: 2011, Focal resection of fast ripples on extraoperative intracranial EEG improves seizure outcome in pediatric epilepsy, *Epilepsia* **52**(10), 1902–1811.

Alarcon, G., Binnie, C., Elwes, R. and Polkey, C.: 1995, Power spectrum and intracranial EEG patterns at seizure onset in partial epilepsy, *Electroencephalography and Clinical Neurophysiology* **94**, 326–337.

Alarcon, G., Garcia Seoane, J., Binnie, C., Miguel, M., Juler, J., Polkey, C., Elwes, R. and Ortiz Blasco, J.: 1997, Origin and propagation of interictal discharges in the acute electrocorticogram. Implications for pathophysiology and surgical treatment of temporal lobe epilepsy, *Brain* **120**, 2259–2282.

An, H. and Gu, L.: 1989, Fast stepwise procedures of selection of variables by using AIC and BIC criteria, *Acta mathematicae applicatae Sinica* **5**(1), 60–67.

Anderson, B. D. O. and Deistler, M.: 1984, Identifiability in dynamic errors-in-variables models, *Journal of Time Series Analysis* **5**, 1–13.

Anderson, T. W.: 2003, *An Introduction to Multivariate Statistical Analysis*, Wiley Series in Probability and Statistics, Wiley.

Andrade-Valenca, L., Dubeau, F., Mari, F., Zelmann, R. and Gotman, J.: 2011, Interictal scalp fast oscillations as markers of the seizure onset zone, *Neurology* **77**, 524–531.

Andrzejak, R., Mormann, F., Widman, G., Kreuz, T., Elger, C. and Lehnertz, K.: 2006, Improved spatial characterization of the epileptic brain by focusing on nonlinearity, *Epileptic Research* **69**, 30–44.

Arroyo, S., Lesser, R., Gordon, B., Fisher, R. and Uematsu, S.: 1993, Subdural Electrodes, *in* E. Niedermeyer and F. Lopes da Silva (eds), *Electroencephalography: Basic Priniciples, Clinical Applications, and Related Fields*, Williams & Wilkins, chapter 40, pp. 695–699.

Astolfi, L., Cincotti, F., Babiloni, C., Carducci, F., Basilisco, A., Rossini, P., Salinari, S., Mattia, D., Cerutti, S., Dayan, B., Ding, L., Ni, Y., He, B. and Babiloni, F.: 2005, Estimation of the cortical connectivity by high-resolution EEG and structural equation modelling: simulations and application to finger tapping data, *IEEE Tansactions on Biomedical Engineering* **52**(5), 757–768.

Astolfi, L., Cincotti, F., Mattia, D., Lai, M., Baccala, L., de Vico Fallani, F., Salinari, S., Ursino, M., Zavaglia, M. and Babiloni, F.: 2005, Comparison of different multivariate methods for the estimation of cortical connectivity: simulations and applications to EEG data, *IEEE EMBC Proceedings 2005* pp. 4484–4487.

Aull-Watschinger, Pataraia, E., Czech, T. and Baumgartner, C.: 2008, Outcome predictors for surgical treatment of temporal lobe epilepsy with hippocampal sclerosis, *Epilepsia* **49**(8), 1308–1316.

Baccala, L. and Sameshima, K.: 2001, Partial directed coherence: a new concept in neural structure determination, *Biological Cybernetics* **84**, 463–474.

Baccala, L., Sameshima, K., Ballester, G., Valle, A. D. and Timo-Iaria, C.: 1998, Studying the Interaction Between Brain Structures via Directed Coherence and Granger Causality, *Applied Signal Processing* **5**, 40–48.

Baccala, L., Takahashi, D. and Sameshima, K.: 2007, Generalized Partial Directed Coherence, *Proceedings of the 15th International Conference on Digital Signal Processing* pp. 163–166.

Bailey, P. and Gibbs, F.: 1951, The surgical treatment of psychomotor epilepsy, *Journal of the American Medical Association* **145**, 365–370.

Baker, G., Jacoby, A., Buck, D., Stalgis, C. and Monnet, D.: 1997, Quality of Life of People with Epilepsy: A European Study, *Epilepsia* **38**(3), 353–362.

Baker, G., Jacoby, A., Smith, D., Dewey, M. and Chadwick, D.: 1994, Development of a Novel Scale to Assess Life Fulfilment as Part of the Further Refinement of a Quality-of-Life Model for Epilepsy, *Epilepsia* **35**(3), 591–596.

Banerjee, P. N. and Hauser, W. A.: 1997, Incidence and prevalence, *in* J. Engel and T. Pedley (eds), *Epilepsy. A comprehensive textbook*, Lippincott Raven, chapter 5, pp. 45–56.

Barbé, K., Pintelon, R. and Schoukens, J.: 2010, Welch Method Revisited. Nonparametric Power Spectrum Estimation Via Circular Overlap, *IEEE Transactions on Signal Processing* **58**(2), 553–565.

Bare, M., Burnstine, T., Fisher, R. and Lesser, R.: 1994, Electroencephalographic Changes During Simple Partial Seizures, *Epilepsia* **35**(4), 715–720.

Barnett, L. and Seth, A.: 2011, Behaviour of Granger causality under filtering: Theoretical invariance and practical application, *Journal of Neuroscience Methods* **201**, 404–419.

Bartholomew, D., Knott, M. and Moustaki, I.: 2011, *Latent Variable Models and Factor Analysis: A Unified Approach*, Wiley Series in Probability and Statistics, Wiley.

Bartlett, M. S.: 1948, Smoothing Periodograms from Time-Series with Continuous Spectra, *Nature* **161**, 686–687.

Bartlett, M. S.: 1950, Periodogram Analysis and Continuous Spectra, *Biometrika* **37**(1/2), 1–16.

Baskind, R. and Birbeck, G.: 2005, Epilepsy-associated stigma in sub-Saharan Africa: The social landscape of a disease, *Epilepsy & Behavior* **7**, 68–73.

Baumgartner, C.: 2001, *Handbuch der Epilepsien*, Springer.

Baumgartner, C., Lindinger, G., Ebner, A., Aull, S., Serles, W., Olbrich, A., Lurger, S., Czech, T., Burgess, R. and Lüders, H.: 1995, Propagation of interictal epileptic activity in temporal lobe epilepsy, *Annals of Neurology* **45**(1), 118–122.

Baumgartner, C. and Pirker, S.: 2012, Presurgical evaluation in adults: noninvasive, *in* H. Stefan and W. Theodore (eds), *Handbook of Clinical Neurology*, Vol. 108, Elsevier, chapter 50, pp. 841–866.

Baumgartner, C., Serles, W., Leutmezer, F., Pataraia, E., Aull, S., Czech, T., Pietrzyk, U., Relic, A. and Podreka, I.: 1998, Preictal SPECT in Temporal Lobe Epilepsy: Regional Cerebral Blood Flow Is Increased Prior to Electroencephalography-Seizure Onset, *Journal of Nuclear Medicine* **39**(6), 978–982.

Behrens, E., Zentner, J., van Roost, D., Hufnagel, A., Elger, C. E. and Schramm, J.: 1994, Subdural and depth electrodes in the presurgical evaluation of epilepsy, *Acta Neurochirurgica* **128**, 84–87.

Berg, A., Berkovic, S., Brodie, M., Buchhalter, J., Cross, J., van Emde Boas, W., Engel, J., French, J., Glauser, T., Mathern, G., Moshé, S., Nordli, D., Plouin, P. and Scheffer, I.: 2010, Revised terminology and concepts for organization of seizures and epilepsies: Report of the ILAE Commission on Classification and Terminology, 2005-2009, *Epilepsia* **51**(4), 676–685.

Berger, H.: 1929, Über das Elektroenkephalogramm des Menschen, *Archiv für Psychiatrie und Nervenkrankheiten* **87**, 527–550.

Blackmann, R. and Tukey, J.: 1958, The Measurement of Power Spectra from the Point of View of Communication Engineering, *Bell System Technical Journal*.

Blanco, J., Stead, M., Krieger, A., Stacey, W., Maus, D., Marsh, E., Viventi, J., Lee, K., Marsh, R., Litt, B. and Worrell, G.: 2011, Data mining neocortical high-frequency oscillations in epilepsy and controls, *Brain* **134**, 2948–2959.

Blanco, J., Stead, M., Krieger, A., Viventi, J., Marsh, W. R., Lee, K., Worrell, G. and Litt, B.: 2010, Unsupervised Classification of High-Frequency Oscillations in Human Neocortical Epilepsy and Control Patients, *Journal of Neurophysiology* **104**, 2900–2912.

Blinowska, K.: 2011, Review of the methods of determination of directed connectivity from multichannel data, *Medical & Biological Engineering & Computing* **49**, 521–529.

Blinowska, K. and Kaminski, M.: 2006, Multivariate Signal Analysis by Parametric Models, *in* B. Schelter, M. Winterhalder and J. Timmer (eds), *Handbook of Time Series Analysis*, Wiley, chapter 15, pp. 373–409.

Blinowska, K., Kus, R., Kaminski, M. and Janiszewska, J.: 2010, Transmission of Brain Activity During Cognitive Task, *Brain Topography* **23**, 205–213.

Blume, W., Borghesi, J. and Lemieux, J.: 1993, Interictal indices of temporal seizure origin, *Annals of Neurology* **34**(5), 703–709.

Blume, W., Holloway, G. and Wiebe, S.: 2001, Temporal Epileptogenesis: Localizing Value of Scalp and Subdural Interictal and Ictal EEG Data, *Epilepsia* **42**(4), 508–514.

Bodenstein, G. and Praetorius, H. M.: 1977, Feature Extraction from the Electroencephalogram by Adaptive Segmentation, *Proceedings of the IEEE* **65**(5), 642–652.

Bodenstein, G., Schneider, W. and Malsburg, C.: 1985, Computerized EEG pattern classification by adaptive segmentation and probability-density-function classification. Description of the method, *Computers in Biology and Medicine* **15**(5), 297–313.

Boudet, S., Peyrodie, L., Gallois, P. and Vasseur, C.: 2007, Filtering by optimal projection and application to automatic artifact removal from EEG, *Signal Processing* **87**, 1978–1992.

Boyden, E., Zhang, F., Bamberg, E., Nagel, G. and Deisseroth, K.: 2005, Millisecond-timescale, genetically targeted optical control of neural acitivity, *Nature Neuroscience* **8**(9), 1263–1268.

Bragin, A., Engel, J. and Staba, R.: 2010, High-frequency oscillations in epileptic brain, *Current Opinion in Neurology* **23**, 151–156.

Brazdil, M., Halamek, J., Jurak, P., Daniel, P., Kuba, R., Chrastina, J., Novak, Z. and Rektor, I.: 2010, Interictal high-frequency oscillations indicate seizure onset zone in patients with focal cortical dysplasia, *Epilepsy Research* **90**, 28–32.

Brillinger, D.: 2001, *Time Series, Data and Analysis*, SIAM.

Brockwell, P. and Davis, R.: 1991, *Time Series: Theory and Methods*, Springer.

Brodie, M., Barry, S., Bamagous, G., Norrie, J. and Kwan, P.: 2012, Patterns of treatment response in newly diagnosed epilepsy, *Neurology* **78**, 1548–1554.

Brodsky, B., Darkhovsky, B., Kaplan, A. and Shishkin, S.: 1999, A nonparametric method for the segmentation of the EEG, *Computer Methods and Programs in Biomedicine* **60**, 93–106.

Burt, C.: 1909, Experimental tests of general intelligence, *British Journal of Psychology* **3**(1-2), 94–177.

Caraballo, R. and Vining, E.: 2012, Ketogenic diet, *in* H. Stefan and W. Theodore (eds), *Handbook of Clinical Neurology*, Vol. 108, Elsevier, chapter 45, pp. 784–793.

Carré, P. and Fernandez-Maloigne, C.: 1998, Research of stationary partitions in nonstationary processes by measurement of spectral distance with the help of nondyadic Malvar's decomposition, *Proceedings of the IEEE International Symposium on Time-Frequency and Time-Scale Analysis* pp. 429–432.

Cascino, G.: 1996, Electroencephalography and Epilepsy, *Journal of Epilepsy* **10**(Supplement 1), 16–23.

Cassidy, M. and Brown, P.: 2002, Hidden Markov based autoregressive analysis of stationary and nonstationary electrophysiological signals for functional coupling studies, *Journal of Neuroscience Methods* **116**, 35–53.

Cattell, R. B.: 1966, The Scree Test For The Number Of Factors, *Multivariate Behavioral Research* **1**(2), 245–276.

Celka, P. and Colditz, P.: 2002, Time-varying statistical dimension analysis with application to newborn scalp EEG seizure signals, *Medical Engineering & Physics* **24**, 1–8.

Chander, R.: 2007, *Algorithms to Detect High Frequency Oscillations in Human Intracerebral EEG*, Master's thesis, McGill University Montreal.

Chatrain, G., Lettich, E. and Nelson, P.: 1985, Ten percent electrode system for topographic studies of spontaneous and evoked EEG activity, *American Journal of EEG Technology* **25**, 83–92.

Chavez, M., Martinerie, J. and Le Van Quyen, M.: 2003, Statistical assessment of nonlinear causality: application to epileptic EEG signals, *Journal of Neuroscience Methods* **124**, 113–128.

Chen, Y., Bressler, S. and Ding, M.: 2006, Frequency decomposition of conditional Granger causality and application to multivariate neural field potential, *Journal of Neuroscience Methods* **150**, 228–237.

Chen, Y., Rangarajan, G., Feng, J. and Ding, M.: 2004, Analyzing multiple nonlinear time series with extended Granger causality, *Physics Letters A* **324**, 26–35.

Chiang, J., Wang, Z. and McKeown, M.: 2009, Sparse multivariate autoregressive (MAR)-based partial directed coherence (PDC) for electroencephalogram (EEG) analysis, *IEEE ICASSP Proceedings 2009* pp. 457–460.

Chu, C.-S. J.: 1995, Time Series Segmentation: A Sliding Window Approach, *Information Sciences* **85**, 147–173.

Clusmann, H., Schramm, J., Kral, T., Helmstaedter, C., Ostertun, B., Fimmers, R., Haun, D. and Elger, C. E.: 2002, Prognostic factors and outcome after different types of resection for temporal lobe epilepsy, *Journal of Neurosurgery* **97**(5), 1131–1141.

Cramer, J., Perrine, K., Devinsky, O. and Meador, K.: 1996, A Brief Questionnaire to Screen for Quality of Life in Epilepsy: The QOLIE-10, *Epilepsia* **37**(6), 577–582.

Crépon, B., Navarro, V., Hasboun, D., Clemenceau, S., Martinerie, J., Baulac, M., Adam, C. and Le Van Quyen, M.: 2010, Mapping interictal oscillations greater than 200 Hz recorded with intracranial macroelectrodes in human epilepsy, *Brain* **133**, 33–45.

Dahlhaus, R.: 1997, Fitting Time Series Models to Nonstationary Processes, *The Annals of Statistics* **25**(1), 1–37.

Dahlhaus, R.: 2000, Graphical interaction models for multivariate time series, *Metrika* **51**, 157–172.

Dahlhaus, R. and Eichler, M.: 2003, Causality and graphical models in time series analysis, *in* P. Green, N. Hjort and S. Richardson (eds), *Highly structured stochastic systems*, Oxford University Press, chapter 1, pp. 115–137.

Daniell, P. J.: 1946, Discussion of "On the Theoretical Specification and Sampling Properties of Autocorrelated Time-Series", *Supplement to the Journal of the Royal Statistical Society* **8**(1), 88–90.

Das, N., Dhanawat, M. and Shrivastava, S.: 2012, An overview on antiepileptic drugs, *Drug Discoveries & Therapeutics* **6**(4), 178–193.

d'Aspremont, A., Ghaoui, L. E., Jordan, M. and Lanckriet, G.: 2007, A direct formulation for sparse PCA using semidefinite programming, *SIAM Review* **49**(3), 434–448.

Davis, R., Lee, T. and Rodriguez-Yam, G.: 2006, Structural Break Estimation for Nonstationary Time Series Models, *Journal of the American Statistical Association* **101**(473), 223–239.

Deistler, M., Anderson, B. D. O., Filler, A., Zinner, C. and Chen, W.: 2010, Generalized Dynamic Factor Models: An Approach via Singular Autoregressions, *European Journal of Control* **16**(3), 211–224.

Deistler, M. and Zinner, C.: 2007, Modelling high-dimensional time series by generalized linear dynamic factor models: an introductory survey, *Communications in Information and Systems* **7**, 153–166.

Delorme, A., Sejnowski, T. and Makeig, S.: 2007, Enhanced detection of artifacts in EEG data using higher-order statistics and independent component analysis, *NeuroImage* **34**, 1443–1449.

Demongeot, J., Bézy-Wendling, J., Mattes, J., Haigron, P., Glade, N. and Coatrieux, J. L.: 2003, Multiscale Modeling and Imaging: The Challenges of Biocomplexity, *Proceedings of the IEEE* **91**(10), 1723 – 1737.

Desmond, J. E., Sum, J. M., Wagner, A. D., Demb, J. B., Shear, P. K., Glover, G. H., Gabrieli, J. D. and Morrell, M. J.: 1995, Functional MRI measurement of language lateralization in Wada-tested patients, *Brain* **118**, 1411–1419.

Dichter, M. and Wilcox, K.: 1997, Excitatory Synaptic Transmission, *in* J. Engel and T. Pedley (eds), *Epilepsy. A comprehensive Texbook*, Lippincott Raven, chapter 23, pp. 251–264.

Dietsch, G.: 1932, Fourier-Analyse von Elektroenkephalogrammen des Menschen, *Pflüger's Archiv für die gesamte Physiologie des Menschen* **230**, 106–112.

Ding, M., Chen, Y. and Bressler, S.: 2006, Granger Cauasality: Basic Theory and Application to Neuroscience, *in* B. Schelter, M. Winterhalder and J. Timmer (eds), *Handbook of Time Series Analysis*, Wiley, chapter 17, pp. 437–460.

Doppelbauer, A., und U. Zifko, J. Z., Baumgartner, C., Mayr, N. and Deecke, L.: 1993, Occurence of epileptiform activity in the routine EEG of epileptic patients, *Acta Neurologica Scandinavica* **87**, 345–352.

Dudek, F., Patrylo, P. and Wuarin, J.: 1999, Mechanisms of neuronal synchronisation during epileptiform activity, *in* A. Degado-Escueta, W. Wilson, R. Olsen and R. Porter (eds), *Jasper's Basic Mechanism of the Epilepsies. Advances in Neurology 79*, Lippincott Williams & Wilkins, pp. 699–706.

Ebersole, J.: 1997, Defining epileptogenic foci: past, present, future, *Journal of Clinical Neurophysiology* **14**(6), 470–483.

Ebersole, J. and Pedley, T. (eds): 2003, *Current Practice of Clinical Electroencephalography*, Lippincott Raven.

Edwards, J.: 2000, *Introduction to graphical modelling*, Springer.

Eichler, M.: 2006a, Graphical modeling of dynamic relationships in multivariate time series, *in* B. Schelter, M. Winterhalder and J. Timmer (eds), *Handbook of Time Series Analysis*, Viley, pp. 335–372.

Eichler, M.: 2006b, On the evaluation of information flow in multivariate systems by the directed transfer function, *Biological Cybernetics* **94**, 469–482.

Eichler, M.: 2007, Granger causality and path diagrams for multivariate time series, *Journal of Econometrics* **137**, 334–353.

Eliasmith, C., Stewart, T., Choo, X., Bekolay, T., DeWolf, T., Tang, C. and Rasmussen, D.: 2012, A Large-Scale Model of the Functioning Brain, *Science* **338**, 1202–1205.

Engel, J.: 1996a, Introduction to temporal lobe epilepsy, *Epilepsy research* **26**, 141–150.

Engel, J.: 1996b, Surgery for seizures, *New England Journal of Medicine* **334**, 647–652.

Engel, J., Rausch, R., Lieb, J., Kuhl, D. and Crandall, P.: 1981, Correlation of Criteria Used for Localizing Epileptic Foci in Patients Considered for Surgical Therapy of Epilepsy, *Annals of Neurology* **9**, 215–224.

Esteller, R., Echauz, J., Tcheng, T., Litt, B. and Pless, B.: 2001, Line length: An efficient feature for seizure onset detection, *Proceedings of the IEEE EMBC* pp. 1707–1710.

Feininger, B.: 2000, Hinfallend Gottes Wort verkünden? Die Epilepsie-Frage im Kontext des Alten Testaments, *in* D. v. Engelhardt, H. Schneble and P. Wolf (eds), *"Das ist eine alte Krankheit". Epilepsie in der Literatur*, Schattauer, pp. 101–122.

Fisher, R., Webber, W., Lesser, R., Arroyo, S. and Uematsu, S.: 1992, High-frequency EEG activitiy at the start of seizures, *Journal of Clinical Neurophysiology* **9**(3), 441–448.

Flamm, C.: 2012, *Novel Methods for Epileptic Onset Zone Detection*, PhD thesis, Vienny University of Technology.

Flamm, C., Graef, A., Pirker, S., Baumgartner, C. and Deistler, M.: 2013, Influence analysis for high-dimensional time series with an application to epileptic seizure onset zone detection, *Journal of Neuroscience Methods* **214**(1), 80–90.

Flamm, C., Kalliauer, U., Deistler, M., Waser, M. and Graef, A.: 2012, Graphs for Dependence and Causality in Multivariate Time Series, *in* L. Wang and H. Garnier (eds), *System Identification, Environmental Modelling, and Control System Design*, Springer, chapter 7, pp. 133–157.

Flink, R., Pedersen, B., Guekht, A., Malmgren, K., Michelucci, R., Neville, B., Pinto, F., Stephani, U. and Özkara, C.: 2002, Guidelines for the use of EEG methodology in the diagnosis of epilepsy, *Acta Neurologica Scandinavica* **106**, 1–7.

Florin, E., Gross, J., Pfeifer, J., Fink, G. R. and Timmermann, L.: 2011, Reliability of multivariate causality measures for neural data, *Journal of Neuroscience Methods* **198**, 344–358.

Foldvary, N., Hammel, J., Bingaman, W., Najm, I. and Lüders, H.: 2001, The localizing value of ictal EEG in focal epilepsy, *Neurology* **57**, 2022–2028.

Formisano, E., Martino, F. D. and Valente, G.: 2008, Multivariate analysis of fMRI time series: classification and regression of brain responses using machine learning, *Magnetic Resonance Imaging* **26**, 921–934.

Franaszczuk, P. and Bergery, G.: 1998, Application of the Directed Transfer Function Method to Mesial and Lateral Onset Temporal Lobe Seizures, *Brain Topography* **11**(1), 13–21.

Franaszczuk, P., Bergey, G. and Kaminski, M.: 1994, Analysis of mesial temporal seizure onset and propagation using the directed transfer function method, *Electroencephalography and Clinical Neurophysiology* **91**, 413–427.

Franaszczuk, P., Blinowska, K. and Kowalczyk, K.: 1985, The Application of Parametric Multichannel Spectral Estimates in the Study of Electrical Brain Activity, *Biological Cybernetics* **51**, 239–247.

Fürbaß, F., Hartmann, M., Perko, H., Skupch, A., Dollfuß, P., Gritsch, G., Baumgartner, C. and Kluge, T.: 2012, Combining Time Series and Frequency Domain Analysis for an Automated Seizure Detection, *Proceedings of the IEEE EMBC* pp. 1020–1023.

French, J. A. and Faught, E.: 2009, Rational polytherapy, *Epilepsia* **50**(suppl. 8), 63–68.

Friedlander, F. G. and Joshi, M.: 1999, *Introduction to the Theory of Distributions*, Cambridge University Press.

Friston, K., Harrison, L. and Penny, W.: 2003, Dynamic Causal Modeling, *NeuroImage* **19**, 1273–1302.

Fujii, M., Inoue, T., Nomura, S., Maruta, Y., He, Y., Koizumi, H., Shirao, S., Owada, Y., Kunitsugu, I., Yamakawa, T., Tokiwa, T., Ishizuka, S., Yamakawa, T. and Suzuki, M.: 2012, Cooling of the epileptic focus suppresses seizures with minimal influence on neurologic functions, *Epilepsia* **53**(3), 485–493.

Fukuda, K., Stanley, H. E. and Amaral, L. N.: 2004, Heuristic segmentation of a nonstationary time series, *Physical Review E* **69**, 021108.

Fusheng, Y., Bo, H. and Qingyu, T.: 2000, Approximate entropy and its application in biosignal analysis, *in* M. Akay (ed.), *Nonlinear Biomedical Signal Processing*, Vol. 2, IEEE Press, chapter 3, pp. 72–91.

Gabor, D.: 1946, Theory of Communication, *Journal of the Institution of Electrical Engineers* **93**(26), 429–457.

Gardner, A., Worrell, G., Marsh, E., Dlugos, D. and Litt, B.: 2007, Human and automated detection of high-frequency oscillations in clinical intracranial EEG recordings, *Clinical Neurophysiology* **118**, 1134–1143.

Gasser, T.: 1979, Statistical Handling of EEG-Data, *Pharmakopsychiatrie* **12**, 210–219.

Gasser, T., Schuller, J. and Schreiter Gasser, U.: 2005, Correction of muscle artefacts in the EEG power spectrum, *Clinical Neurophysiology* **116**, 2044–2050.

Gath, I., Feuerstein, C., Pham, D. and Rondouin, G.: 1992, On the Tracking of Rapid Dynamic Changes in Seizure EEG, *IEEE Transactions on Biomedical Engineering* **39**(9), 952–958.

Ge, M., Jiang, X., Bai, Q., Yang, S., Gusphyl, J. and Yan, W.: 2007, Application of the Directed Transfer Function Method to the Study of the Propagation of Epilepsy Neural Information, *Proceedings of the IEEE EMBC* pp. 3266–3269.

Geng, S. and Zhou, W.: 2010, Nonlinear Feature Comparison of EEG Using Correlation Dimension and Approximate Entropy, *Proceedings of the IEEE BMEI* pp. 978–981.

Geweke, J.: 1977, The Dynamic Factor Analysis of Economic Time Series Models, *in* Aigner and Goldberger (eds), *Latent Variables in Socio-Economic Models*, North-Holland, chapter 19, pp. 365–383.

Geweke, J.: 1982, Measurement of Linear Dependence and Feedback between Multiple Time Series, *Journal of the American Statistical Association* **77**, 304–313.

Geweke, J.: 1984, Measurement of Conditional Linear Dependence and Feedback between Multiple Time Series, *Journal of the American Statistical Association* **79**, 907–915.

Gilliam, F., Barry, J., Hermann, B., Meador, K., Vahle, V. and Kanner, A.: 2006, Rapid detection of major depression in epilepsy: a multicentre study, *Lancet Neurology* **5**, 399–405.

Ginter, J., Blinowska, K., Kaminski, M. and Durka, P.: 2001, Phase and amplitude analysis in time-frequency space – application to voluntary finger movement, *Journal of Neuroscience Methods* **110**, 113–124.

Gotman, J.: 2003, Automatic Detection and Analysis of Seizures and Spikes, *in* J. Ebersole and T. Pedley (eds), *Current Practice of Clinical Electroencephalography*, Lippincott Raven, chapter 22, pp. 713–731.

Gotman, J. and Koffler, D.: 1989, Interictal spiking increases after seizures but does not after decrease in medication, *Electroencephalography and Clinical Neurophysiology* **72**(1), 7–15.

Gourine, A., Kasymov, V., Marina, N., Tang, F., Figueiredo, M., Lane, S., Teschemacher, A., Spyer, K., Deisseroth, K. and Kasparov, S.: 2010, Astrocytes Control Breathing Through ph-Dependent Release of ATP, *Science* **329**, 571–575.

Gourévitch, B., Le Bouquin-Jeannès, R. and Faucon, G.: 2006, Linear and nonlinear causality between signals: methods, examples and neurophysiological applications, *Biological Cybernetics* **95**, 349–369.

Graef, A.: 2008, *Nonstationary Autoregressive Modeling for Epileptic Seizure Propagation Analysis*, Master's thesis, Vienna University of Technology.

Graef, A.: 2013, *Automated Procedures for Focus Detection and Seizure Propagation Analysis from Ictal ECoG*, PhD thesis, Vienny University of Technology.

Graef, A., Flamm, C., Pirker, S., Baumgartner, C., Deistler, M. and Matz, G.: 2013, Automatic ictal HFO detection for determination of initial seizure spread, *Proceedings of the IEEE EMBC* pp. 2096–2099.

Graef, A., Flamm, C., Pirker, S., Deistler, M. and Baumgartner, C.: 2012, A physiologically motivated ECoG segmentation method for epileptic seizure onset zone detection, *Proceedings of the IEEE EMBC* pp. 3500–3503.

Graef, A., Hartmann, M., Deistler, M. and Kluge, T.: 2009, Regression-based analysis of synchronization in multichannel EEG in epilepsy, *Proceedings of the IEEE EMBC* pp. 4743–4746.

Graef, A., Hartmann, M., Flamm, C., Baumgartner, C., Deistler, M. and Kluge, T.: 2013, A novel method for the identification of synchronization effects in multichannel ECoG with an application to epilepsy, *Biological Cybernetics* **107**(3), 321–335.

Granger, C.: 1969, Investigating causal relations by econometric models and cross-spectral methods, *Econometrica* **37**, 424–438.

Grech, R., Cassar, T., Muscat, J., Camilleri, K., Fabri, S., Zervakis, M., Xanthopoulos, P., Sakkalis, V. and Vanrumste, B.: 2008, Review on solving the inverse problem in EEG source analysis, *Journal of Neuro-Engineering and Rehabilitation* **5**.

Gregory, R., Oates, T. and Merry, R.: 1993, Electroencephalogram epileptiform abnormalities in candidates for aircrew training, *Electroencephalography and Clinical Neurophysiology* **86**, 75–77.

Grossmann, A. (ed.): 2013, *Pschyrembel Klinisches Wörterbuch*, de Gruyter.

Götz-Trabert, K., Hauck, C., Wagner, K., Fauser, S. and Schulze-Bonhage, A.: 2008, Spread of ictal activitiy in focal epilepsy, *Epilepsia* **49**(9), 1594–1601.

Guangming, Z., Huancong, Z., Wenjing, Z., Guoqiang, C., Dongming, W., Yanfang, S., Xiaohua, L. and Jiuluan, L.: 2009, Synchronous recording of intracranial and extracranial EEG in temporal lobe epilepsy, *Epilepsy Research* **85**, 46–52.

Guldvogl, B., Loyning, Y., Hauglie-Hanssen, E., Flood, S. and Bjonaes, H.: 1994, Predictive Factors for Success in Surgical Treatment for Partial Epilepsy: A Multivariate Analysis, *Epilepsia* **35**(3), 566–578.

Guo, S., Seth, A., Kendrick, K., Zhou, C. and Feng, J.: 2008, Partial Granger causality - eliminationg exogenous inputs and latent variables, *Journal of Neuroscience Methods* **172**, 79–93.

Hallez, H., De Vos, M., Vanrumste, B., Van Hese, P., Assecondi, S., Van Laere, K., Dupont, P., Van Paesschen, W., Van Huffel, S. and Lemahieu,

I.: 2009, Removing muscle and eye artifacts using blind source separation techniques in ictal EEG source imaging, *Clinical Neurophysiology* **120**, 1262–1272.

Hanada, T., Hashizume, Y., Tokuhara, N., Takenaka, O., Kohmura, N., Ogasawara, A., Hatakeyama, S., amd Masataka Ueno, M. O. and Nishizawa, Y.: 2011, Perampanel: A novel, orally active, noncompetitive AMPA-receptor antagonist that reduces seizure activity in rodent models of epilepsy, *Epilepsia* **52**(7), 1331–1340.

Hannan, E. and Deistler, M.: 2012, *The statistical theory of linear systems*, Classics in Applied Mathematics, SIAM.

Hartmann, M., Fürbaß, F., Perko, H., Skupch, A., Lackmayer, K., Baumgartner, C. and Kluge, T.: 2011, EpiScan: Online seizure detection for epilepsy monitoring units, *Proceedings of the IEEE EMBC* pp. 6096–6099.

Hartmann, M., Graef, A., Perko, H., Baumgartner, C. and Kluge, T.: 2008, A Novel Method for the Characterization of Synchronization and Coupling in Multichannel EEG and ECoG, *WASET Proceedings 2008* **34**, 6–11.

Hashiguchi, K., Morioka, T., Yoshida, F., Miyagi, Y., Nagata, S., Sakata, A. and Sasaki, T.: 2007, Correlation between scalp-recorded electroencephalographic and electrocorticographic activities during ictal period, *Seizure* **16**, 238–247.

Hauser, W., Andereson, V., Loewenson, R. and McRoberts, S.: 1982, Seizure recurrence after a first unprovoked seizure, *New England Journal of Medicine* **307**(9), 522–528.

He, P., Wilson, G. and Russel, C.: 2004, Removal of ocular artifacts from electro-encephalogram by adaptive filtering, *Medical & Biological Engineering & Computing* **42**, 407–412.

Hegde, A., Erdogmus, D., Shiau, D. S., Principe, J. and Sackellares, C.: 2005, Quantifying spatio-temporal dependencies in epileptic ECoG, *Signal Processing* **85**, 2082–2100.

Henry, T., Ross, D., Shuh, L. and Drury, I.: 1999, Indications and Outcome of Ictal Recording With Intracerebral and Subdural Electrodes in Refractory Complex Partial Seizures, *Journal of Clinical Neurophysiology* **16**(5), 426–438.

Hermann, B., Seidenberg, M. and Bell, B.: 2000, Psychiatric Comorbidity in Chronic Epilepsy: Identification, Consequences, and Treatment of Major Depression, *Epilepsia* **41**(Suppl 2), S31–S41.

Herrmann, C. S. and Demiralp, T.: 2005, Human EEG gamma oscillations in neuropsychiatric disorders, *Clinical Neurophysiology* **116**, 2719–2733.

Hesse, W., Mller, E., Arnold, M. and Schack, B.: 2003, The use of time-variant EEG Granger causality for inspecting directed interdependencies of neural assemblies, *Journal of Neuroscience Methods* **124**, 27–44.

Hilbert, D.: 1912, *Grundzüge einer allgemeinen Theorie der linearen Integralgleichungen*, Teubner.

Hirtz, D., Thurman, D., Gwinn-Hardy, K., Mohamed, M., Chaudhuri, A. and Zalutsky, R.: 2007, How common are the "common" neurologic disorders?, *Neurology* **68**, 326–337.

Hodgkin, A. L. and Huxley, A. F.: 1952, A quantitative description of membrane current and its application to conduction and excitation in nerve, *Journal of Physiology* **117**, 500–544.

Hotelling, H.: 1933, Analysis of a complex of statistical variables into principal components., *Journal of Educational Psychology* **26**(6), 417–441.

Hughlings-Jackson, J.: 1873, On the anatomical, physiological, and pathological investigation of epilepsies, *West Riding Lunatic Asylum Reports* **3**, 315.

Imamura, H., Matsumoto, R., Inouchi, M., Matsuhashi, M., Mikuni, N., Takahashi, R. and Ikeda, A.: 2011, Ictal wideband ECoG: Direct comparison between ictal slow shifts and high frequency oscillations, *Clinical Neurophysiology* **122**, 1500–1504.

Inoue, T., Kobayashi, K., Oka, M., Yoshinaga, H. and Ohtsuka, Y.: 2008, Spectral characteristics of EEG gamma rhythms associated with epileptic spasms, *Brain & Development* **30**, 321–328.

Inouye, T., Toi, S. and Matsumoto, Y.: 1995, A new segmentation method of electroencephalograms by use of Akaike's information criterion, *Cognitive Brain Research* **3**, 33–40.

Jackson, J.: 2004, *A User's Guide to Principal Components*, John Wiley & Sons.

Jackson, J. D.: 1998, *Classical Electrodynamics*, John Wiley & Sons.

Jacobs, J., LeVan, P., Chander, R., Hall, J., Dubeau, F. and Gotman, J.: 2008, Interictal high-frequency oscillations (80-500 Hz) are an indicator of seizure onset areas independent of spikes in the human epileptic brain, *Epilepsia* **49**(11), 1893–1907.

Jacobs, J., Staba, R., Asano, E., Otsubo, H., Wu, J. Y., Zijlmans, M., Mohamed, I., Kahane, P., Dubeau, F., Navarro, V. and Gotman, J.: 2012, High-frequency oscillations (HFOs) in clinical epilepsy, *Progress in Neurobiology* **98**(3), 302–315.

Jacobs, J., Zelmann, R., Jirsch, J., Chander, R., Chatillon, C.-E., Dubeau, F. and Gotman, J.: 2009, High frequency oscillations (80-500 Hz) in the preictal period in patients with focal seizures, *Epilepsia* **50**(7), 1780–1792.

Jacobs, J., Zijlmans, M., Zelmann, R., Chatillon, C.-E., Hall, J., Olivier, A., Dubeau, F. and Gotman, J.: 2010, High-Frequency Electroencephalographic Oscillations Correlate With Outcome of Epilepsy Surgery, *Annals of Neurology* **67**(2), 209–220.

James, C. and Gibson, O.: 2003, Temporally constrained ICA: An Application to Artifact Rejection in Electromagnetic Brain Signal Analysis, *IEEE Transactions on Biomedical Engineering* **50**(9), 1108–1116.

Janszky, J., Fogarasi, A., Jokeit, H., Schulz, R., Hoppe, M. and Ebner, A.: 2001, Spatiotemporal relationship between seizure activity and interictal spikes in temporal lobe epilepsy, *Epilepsy Research* **47**, 179–188.

Jasper, H.: 1958, The ten-twenty electrode system of the International Federation, *Electroencephalography and Clinical Neurophysiology* **10**, 371–375.

Jefferys, J., de la Prida, L. M., Wendling, F., Bragin, A., Avoli, M., Timofeev, I. and da Silva, F. L.: 2012, Mechanisms of physiological and epileptic HFO generation, *Progress in Neurobiology* **98**, 250–264.

Jenkins, G. and Watts, D.: 1968, *Spectral analysis and its applications*, Holden-Day.

Jenssen, S., Roberts, C., Gracely, E., Dlugos, D. and Sperling, M.: 2011, Focal seizure propagation in the intracranial EEG, *Epilepsy Research* **93**, 25–32.

Jirsch, J., Urrestarazu, E., LeVan, P., Olivier, A., Dubeau, F. and Gotman, J.: 2006, High-frequency oscillations during human focal seizures, *Brain* **129**, 1593–1608.

Jiruska, P., Powell, A., Chang, W.-C. and Jefferys, J.: 2010, Electrographic high-frequency activity and epilepsy, *Epilepsy Research* **89**, 60–65.

Jouny, C., Bergey, G. and Franaszczuk, P.: 2010, Partial seizures are associated with an early increase in signal complexity, *Clinical Neurophysiology* **121**, 7–13.

Jouny, C., Franaszczuk, P. and Bergey, G.: 2003, Characterization of epileptic seizure dynamics using Gabor atom density, *Clinical Neurophysiology* **114**, 426–437.

Joyce, C., Gorodnitsky, I. and Kutas, M.: 2004, Automatic removal of eye movement and blink artifacts from EEG data using blind component separation, *Psychophysiology* **41**, 313–325.

Jung, T.-P., Humphries, C., Lee, T.-W., Makeig, S., McKeown, M., Iragui, V. and Sejnowski, T.: 1998, Extended ICA Removes Artifacts from Electroencephalograhic Recordings, *Advances in Neural Information Processing Systems* **10**, 894–900.

Jung, T.-P., Makeig, S., Westerfield, M., Townsend, J., Courchesne, E. and Sejnowski, T.: 2000, Removal of eye activity artifacts from visual event-related potentials in normal and clinical subjects, *Clinical Neurophysiology* **111**, 1745–1758.

Kahane, P. and Spencer, S.: 2012, Invasive evaluation, *in* H. Stefan and W. Theodore (eds), *Handbook of Clinical Neurology*, Vol. 108, Elsevier, chapter 51, pp. 868–879.

Kaiser, J.: 1990, On a simple algorithm to calculate the 'energy' of a signal, *Proceedings of the IEEE ICASSP* pp. 381–384.

Kaminski, M. and Blinowska, K.: 1991, A new method of the description of the information flow in the brain structures, *Biological Cybernetics* **65**, 203–210.

Kaminski, M., Ding, M., Truccolo, W. and Bressler, S.: 2001, Evaluating causal relations in neural systems: Granger causality, directed transfer function and statistical assessment of significance, *Biological Cybernetics* **85**, 145–157.

Kandel, E., Schwartz, J., Jessell, T., Siegelbaum, S. and Hudspeth, A. (eds): 2012, *Principles of Neural Science*, McGraw-Hill.

Kanemoto, K., Kawasaki, J. and Kawai, I.: 1997, The Lateralizing Value of Scalp Ictal EEG Patterns in Temporal Lobe Epilepsy with Unilateral Hippocampal Atrophy with Special Attention to the Initial Slow Waves, *Journal of Epilepsy* **10**, 225–231.

Kaplan, A., Fingelkurts, A., Fingelkurts, A., Borisov, S. and Darkhovsky, B.: 2005, Nonstationary nature of the brain activitiy as revealed by EEG/EMG: Methodological, practical and conceptual challenges, *Signal Processing* **85**, 2190–2212.

Kaplan, A., Röschke, J., Darkhovsky, B. and Fell, J.: 2001, Macrostructural EEG characterization based on nonparametric change point segmentation: application to sleep analysis, *Journal of Neuroscience Methods* **106**, 81–90.

Karlin, S. and Rubin, H.: 1956, The Theory of Decision Procedures for Distributions with Monotone Likelihood Ratio, *Annals of Mathematical Statistics* **27**(2), 272–299.

Kay, S. M.: 1998, *Fundamentals of Statistical Signal Processing. Volume II: Detection Theory*, Prentice Hall.

Keogh, E., Chu, S., Hart, D. and Pazzani, M.: 2004, Segmenting Time Series: A Survey and Novel Approach, *in* M. Last, A. Kandel and H. Bunke (eds), *Data mining in time series databases*, Vol. 57 of *Machine Perception Artificial Intelligence*, World Scientific, chapter 1, pp. 1–21.

Khosravani, H., Mehrotra, N., Rigby, M., Hader, W., Pinnegar, C. R., Pillay, N., Wiebe, S. and Federico, P.: 2009, Spatial localization and time-dependent changes of electrographic high frequency oscillations in human temporal lobe epilepsy, *Epilepsia* **50**(4), 605–616.

Kilga, M.: 1993, *Spektralschätzung*, Master's thesis, Vienna University of Technology.

Kim, J. S., Im, C. H., Jung, Y. J., Kim, E. Y., Lee, S. K. and Chung, C. K.: 2010, Localization and propagation analysis of ictal source rhythm by electrocorticography, *NeuroImage* **52**, 1279–1288.

King, M., Newton, M., Jackson, G., Fitt, G., Mitchell, L., Silvapulle, M. and Berkovic, S.: 1998, Epileptology of the first-seizure presentation: a clinical, electroencephalographic, and magnetic resonance imaging study of 300 consecutive patients, *Lancet* **352**, 1007–1011.

Kipinski, L., König, R., Sieluzycki, C. and Kordecki, W.: 2011, Application of modern tests for stationarity to single-trial MEG data. Transferring powerful statistical tools from econometrics to neuroscience, *Biological Cybernetics* **105**, 183–195.

Kobayashi, K., Agari, T., Oka, M., Yoshinaga, H., Date, I., Ohtsuka, Y. and Gotman, J.: 2010, Detection of seizure-associated high-frequency oscillations above 500 Hz, *Epilepsy Research* **88**, 139–144.

Kobayashi, K., Inoue, T., Watanabe, Y., Oka, M., Endoh, F., Yoshinaga, H. and Ohtsuka, Y.: 2009, Spectral analysis of EEG gamma rhythms associated with tonic seizures in Lennox-Gastaut syndrome, *Epilepsy Research* **86**, 15–22.

Kong, X., Lou, X. and Thakor, N.: 1997, Detection of EEG Changes via a Generalized Itakura Distance, *Proceedings of the IEEE EMBC* pp. 1540–1542.

Kong, X., Thakor, N. and Goel, V.: 1995, Characterization of EEG signal changes via Itakura distance, *Proceedings of the IEEE EMBC and CMBEC* pp. 783–874.

Korzeniewska, A., Manczak, M., Kaminski, M., Blinowska, K. and Kasicki, S.: 2003, Determination of information flow direction among brain structures by a modified directed transfer function (dDTF) method, *Journal of Neuroscience Methods* **125**, 195–207.

Kus, R., Kaminski, M. and Blinowska, K.: 2004, Determination of EEG Activity Propagation: Pair-Wise Versus Multichannel Estimate, *IEEE Transactions on Biomedical Engineering* **51**(9), 1501–1510.

Kwan, P., Arzimanoglou, A., Berg, A. T., Brodie, M., Hauser, W. A., Mathern, G., Moshé, S., Perucca, E., Wiebe, S. and French, J.: 2010, Definition of drug resistant epilepsy: Consensus proposal by the ad hoc Task Force of the ILAE Commission on Theraupeutic Strategies, *Epilepsia* **51**(6), 1069–1077.

Kwan, P. and Brodie, M.: 2000, Early identification of refractory epilepsy, *New England Journal of Medicine* **342**, 314–319.

Kwan, P. and Brodie, M.: 2001, Effectiveness of First Antiepileptic Drug, *Epilepsia* **42**(10), 1255–1260.

Kwan, P., Schachter, S. and Brodie, M.: 2011, Drug-Resistant Epilepsy, *New England Journal of Medicine* **365**(10), 919–926.

Lai, V., Mak, H., Yung, A., Ho, W. and Hung, K.: 2010, Neuroimaging techniques in epilepsy, *Hong Kong Medical Journal* **16**(4), 292–298.

Lantz, G., Grave de Peralta, R., Spinelli, L., Seeck, M. and Michel, C. M.: 2003, Epileptic source localization with high density EEG: how many electrodes are needed?, *Clinical Neurophysiology* **114**, 63–69.

Lavielle, M.: 1993, Detection of Changes in the Spectrum of a Multidimensional Process, *IEEE Transactions on Signal Processing* **41**(2), 742–749.

Lavielle, M.: 1998, Optimal Segmentation of Random Processes, *IEEE Transactions on Signal Processing* **46**(5), 1365–1373.

Lüders, H. (ed.): 1991, *Epilepsy Surgery*, Raven Press.

Lüders, H. and Noachtar, S.: 1994, *Atlas und Klassifikation der Elektroenzephalographie: Einführung in die EEG-Auswertung*, Ciba-Geigy Verlag.

Lüders, H. O. and Awad, I.: 1991, Conceptual considerations, *in* H. Lüders (ed.), *Epilepsy Surgery*, Raven Press, chapter 7, pp. 51–62.

Lee, S.-A., Yim, S., Lim, Y., Kang, J. and Lee, J.: 2006, Factors predicting seizure outcome of anterior temporal lobectomy for patients with mesial temporal sclerosis, *Seizure* **15**, 397–404.

Lehnertz, K. and Elger, C.: 1995, Spatio-temporal dynamics of the primary epileptogenic area in temporal lobe epilepsy characterized by neuronal complexity loss, *Electroencephalography and Clinical Neurophysiology* **95**, 108–117.

Lesser, R., Crone, N. and Webber, W.: 2010, Subdural electrodes, *Clinical Neurophysiology* **121**, 1376–1392.

LeVan, P., Urrestarazu, E. and Gotman, J.: 2006, A system for automatic artifact removal in ictal scalp EEG based on independent component analysis and Bayesian classification, *Clinical Neurophysiology* **117**, 912–927.

Liao, W., Mantini, D., Zhang, Z., Pan, Z., Ding, J., Gong, Q., Yang, Y. and Chen, H.: 2010, Evaluating the effective connectivity of resting state networks using conditional Granger causality, *Biological Cybernetics* **102**, 57–69.

Lieb, J., Engel, J. and Babb, T.: 1986, Interhemispheric Propagation Time of Human Hippocampal Seizures I. Relationship to Surgical Outcome, *Epilepsia* **27**(3), 286–293.

Lieb, J., Walsh, G., Babb, T., Walter, R. and Crandall, P.: 1976, A Comparison of EEG Seizure Patterns Recorded with Surface and Depth Electrodes in Patients with Temporal Lobe Epilepsy, *Epilepsia* **17**, 137–160.

Lindgren, G.: 2012, *Stationary Stochastic Processes. Theory and Applications*, CRC Press.

Liu, T. and Yao, D.: 2006, Removal of the ocular artifacts from EEG data using a cascade spatio-temporal processing, *Computer Methods and Programs in Biomedicine* **83**, 95–103.

Loehlin, J.: 2004, *Latent Variable Models: An Introduction to Factor, Path, and Structural Equation Analysis*, Lawrence Erlbaum Associates.

Lopatka, M., Laplanche, C., Adam, O., Motsch, J.-F. and Zarzycki, J.: 2005, Non-Stationary Time-Series Segmentation Based on the Schur Prediction Error Analysis, *Proceedings of the IEEE/SP 13th Workshop on Statistical Signal Processing* pp. 251–255.

Lütkepohl, H.: 2007, *New Introduction to Multiple Time Series Analysis*, Springer.

Macdonald, R.: 1997, Inhibitory Synaptic Transmission, *in* J. Engel and T. Pedley (eds), *Epilepsy. A comprehensive Texbook*, Lippincott Raven, chapter 24, pp. 265–276.

Maiwald, T., Mammen, E., Nandi, S. and Timmer, J.: 2008, Surrogate Data - A Qualitative and Quantitative Analysis, *in* R. Dahlhaus, J. Kurths, P. Maass and J. Timmer (eds), *Mathematical Methods in Time Series Analysis and Digital Image Processing*, Springer, chapter 2, pp. 41–74.

Malmivuo, J. and Plonsey, R.: 1995, *Bioelectromagnetism: Principles and Applications of Bioelectric and Biomagnetic Fields*, Online edition (www.bem.fi/book), Oxford University Press.

Mantini, D., Perrucci, M., Cugini, S., Ferretti, A., Romani, G. and Del Gratta, C.: 2007, Complete artifact removal for EEG recorded during continuous fMRI using independent component analysis, *NeuroImage* **34**, 598–607.

Marinazzo, D., Liao, W., Chen, H. and Stramaglia, S.: 2011, Non-linear connectivity by Granger causality, *Neuroimage* **58**, 330–338.

Marple, S.: 1987, *Digital spectral analysis with applications*, Prentice Hall.

May, T. and Pfäfflin, M.: 2001, Evaluating comprehensive care: Description of the PESOS and its psychometric properties, *in* M. Pfäfflin, R. Fraser, R. Thorbecke, U. Specht and P. Wolf (eds), *Comprehensive Care for People with Epilepsy*, John Libbey & Company, chapter 35, pp. 319–339.

Michel, C. M., Lantz, G., Spinelli, L., de Peralta, R. G., Landis, T. and Seeck, M.: 2004, 128-Channel EEG Source Imaging in Epilepsy: Clinical Yield and Localization Precision, *Journal of Clinical Neurophysiology* **21**(2), 71–83.

Mitra, S.: 2002, *Digital Signal Processing. A computer-based approach*, McGraw-Hill.

Mizuno-Matsumoto, Y., Okazaki, K., Kato, A., Yoshimine, T., Sato, Y., Tamura, S. and Hayakawa, T.: 1999, Visualization of Epileptogenic Phenomena Using Cross-Correlation Analysis: Localization of Epileptic Foci and Propagation of Epileptiform Discharges, *IEEE Transactions on Biomedical Engineering* **46**(3), 271–279.

Molenaar, P. C.: 1985, A dynamic factor model for the analysis of multivariate timeseries, *Psychometrika* **50**(2), 181–202.

Molenaar, P. C. and Nesselroade, J. R.: 2001, Rotation in the dynamic factor modelling of multivariate stationary time series, *Psychometrika* **66**(1), 99–107.

Mormann, F., Andrzejak, R., Elger, C. E. and Lehnertz, K.: 2007, Seizure prediction: the long and winding road, *Brain* **130**, 314–333.

Mormann, F., Kreuz, T., Andrzejak, R., David, P., Lehnertz, K. and Elger, C.: 2003, Epileptic seizures are preceded by a decrease in synchronization, *Epilepsy Research* **53**, 173–185.

Nelson, R., Myers, S., Simonotto, J., Furmann, M., Spano, M., Norman, W., Liu, Z., DeMarse, T., Carney, P. and Ditto, W.: 2006, Detection of High Frequency Oscillations with Teager Energy in an Animal Model of Limbic Epilepsy, *Proceedings of the IEEE EMBC* pp. 2578–2580.

Netoff, T. and Schiff, S.: 2002, Decreased Neuronal Synchronization during Experimental Seizures, *Journal of Neuroscience* **22**(16), 7297–7307.

Neyman, J. and Pearson, E.: 1933, On the Problem of the Most Efficient Tests of Statistical Hypotheses, *Philosophical Transactions of the Royal Society of London. Series A, Mathematical, Physical & Engineering Sciences* **231**, 289–337.

Niedermeyer, E. and Lopes da Silva, F. (eds): 1993, *Electroencephalography: Basic Principles, Clinical Applications, and Related Fields*, Williams & Wilkins.

Noachtar, S., Binnie, C., Ebersole, J., Mauguière, F., Sakamoto, A. and Westmoreland, B.: 1999, A glossary of terms most commonly used by the clinical electroencephalographers and proposal for the report form for the EEG findings, *Electroencephalography and Clinical Neurophysiology* **Suppl. 52**, 21–41.

Noachtar, S., Binnie, C., Ebersole, J., Mauguière, F., Sakamoto, A. and Westmoreland, B.: 2004, Glossar der meistgebrauchten Begriffe in der klinischen Elektroenzephalographie und Vorschläge für die EEG-Befunderstellung, *Klinische Neurophysiologie* **35**, 5–21.

Noachtar, S., Rosenow, F., Arnold, S., Baumgartner, C., Ebner, A., Hamer, H., Holthausen, H., Meencke, H., Müller, A., Sakamoto, A., Steinhoff, B., Tuxhorn, I., Werhahn, K., Winkler, P. and Lüders, H.: 1998, Die semiologische Klassifikation epileptischer Anfälle, *Nervenarzt* **69**, 117–126.

Nolan, H., Whelan, R. and Reilly, R.: 2010, FASTER: Fully Automated Statistical Thresholding for EEG artifact Rejection, *Journal of Neuroscience Methods* **192**, 152–162.

Ohlsson, H., Ljung, L. and Boyd, S.: 2010, Segmentation of ARX-models using sum-of-norms regularization, *Automatica* **46**, 1107–1111.

Ombao, H., Raz, J., von Sachs, R. and Malow, B.: 2001, Automatic Statistical Analysis of Bivariate Nonstationary Time Series, *Journal of the American Statistical Association* **96**(454), 543–560.

Ombao, H., von Sachs, R. and Guo, W.: 2005, SLEX Analysis of Multivariate Nonstationary Time Series, *Journal of the American Statistical Association* **100**(470), 519–531.

Omidvarnia, A., Azemi, G., Boashash, B., O'Toole, J., Colditz, P. and Vanhatalo, S.: 2012, Orthogonalized Partial Directed Coherence for Functional Connectivity Analysis of Newborn EEG, *Proceedings of the 19th International Conference on Neural Information Processing* pp. 683–691.

Oppenheim, A. and Schafer, R.: 1989, *Discrete-Time Signal Processing*, Prentice Hall.

Ostenveld, R. and Praamstra, P.: 2001, The five percent electrode system for high-resolution EEG and ERP measurements, *Clinical Neurophysiology* **112**, 713–719.

Ozaki, I. and Hashimoto, I.: 2011, Exploring the physiology and function of high-frequency oscillations (HFOs) from the somatosensory cortex, *Clinical Neurophysiology* **122**, 1908–1923.

Pacia, S. and Ebersole, J.: 1997, Intracranial EEG Substrates of Scalp Ictal Patterns from Temporal Lobe Foci, *Epilepsia* **38**(6), 642–654.

Pacia, S. and Ebersole, J.: 1999, Intracranial EEG in Temporal Lobe Epilepsy, *Journal of Clinical Neurophysiology* **16**(5), 399–407.

Panayiotopoulos, C. P.: 2011, *Principles of Therapy in the Epilepsies*, Springer.

Papana, A. and Kugiumtzis, D.: 2012, Detection of direct causal effects and application to epileptic encephalogram analysis, *International Journal of Bifurcation and Chaos* **22**(9), 1250222.

Paulsen, F. and Waschke, J. (eds): 2010, *Sobotta. Atlas der Anatomie des Menschen. Kopf, Hals und Neuroanatomie*, Urban & Fischer.

Pearl, J.: 2000, *Causality*, Cambridge University Press.

Pearson, K.: 1901, On lines and planes of closest fit to systems of points in space, *Philosophical Magazine Series 6* **2**(11), 559–572.

Penfield, W.: 1934, Surgery in the Treatment of Epilepsy, *Surgery, Gynecology & Obstetrics* **58**, 1041.

Penm, J. and Terrell, R.: 1982, On the recursive fitting of subset autoregressions, *Journal of Timeseries Analysis* **3**(1), 43–59.

Penny, W. and Roberts, S.: 1999, Dynamic models for nonstationary signal segmentation, *Computers and Biomedical Research* **32**, 483–502.

Pereda, E., Quiroga, R. Q. and Bhattacharya, J.: 2005, Nonlinear multivariate analysis of neurophysiological signals, *Progress in Neurobiology* **77**, 1–37.

Perucca, E. and Tomson, T.: 2011, The pharmacological treatment of epilepsy in adults, *Lancet Neurology* **10**, 446–456.

Picone, J. W.: 1993, Signal modeling techniques in speech recognition, *Proceedings of the IEEE* **81**(9), 1215–1247.

Plummer, C., Harvey, A. and Cook, M.: 2008, EEG source localization in focal epilepsy: Where are we now?, *Epilepsia* **49**(2), 201–218.

Pondal-Sordo, M., Diosy, D., Téllez-Zenteno, J. F., Sahjpaul, R. and Wiebe, S.: 2007, Usefulness of intracranial EEG in the decision process for epilepsy surgery, *Epilepsy Research* **74**, 176–182.

Porter, R., Dhir, A., Macdonald, R. and Rogawski, M.: 2012, Mechanisms of action of antiseizure drugs, *in* H. Stefan and W. Theodore (eds), *Handbook of Clinical Neurology*, Vol. 108, Elsevier, chapter 39, pp. 664–681.

Prado, R., West, M. and Krystal, A.: 2001, Multichannel electrocencephalographic analyses via dynamic regression models with time-varying lag-lead structure, *Journal of the Royal Statistical Society Series C* **50**(1), 95–109.

Praetorius, H. M., Bodenstein, G. and Creutzfeldt, O. D.: 1977, Adaptive segmentation of EEG records: A new approach to automatic EEG analysis, *Electroencephalography and Clinical Neurophysiology* **42**, 84–94.

Priestley, M.: 1965, Evolutionary Spectra and Non-stationary Processes, *Journal of the Royal Statistical Society. Series B* **2**, 204–237.

Puthusserypady, S. and Ratnarajah, T.: 2006, Robust adaptive techniques for minimization of EOG artefacts from EEG signals, *Signal Processing* **86**, 2351–2363.

Quesney, L. and Niedermeyer, E.: 1993, Electrocorticography, *in* E. Niedermeyer and F. Lopes da Silva (eds), *Electroencephalography: Basic Priniciples, Clinical Applications, and Related Fields*, Williams & Wilkins, chapter 40, pp. 695–699.

Randall, R. and Tech, B.: 1987, *Frequency analysis*, Brüel & Kjaer.

Ray, A., Tao, J., Hawes-Ebersole, S. and Ebersole, J.: 2007, Localizing value of scalp EEG spikes: A simultaneous scalp and intracranial study, *Clinical Neurophysiology* **118**, 69–79.

Ren, L., Terada, K., Baba, K., Usui, N., Umeoka, S., Usui, K., Matsuda, K., Tottori, T., Nakamura, F., Mihara, T. and Inoue, Y.: 2011, Ictal Very Low Frequency Oscillations in Human Epilepsy Patients, *Annals of Neurology* **69**(1), 201–206.

Richardson, M.: 2011, New observations may inform seizure models: Very fast and very slow oscillations, *Progress in Biophysics and Molecular Biology* **105**, 5–13.

Riesz, M.: 1928, Sur les fonctions conjugées, *Mathematische Zeitschrift* **27**(1), 218–244.

Risinger, M., Engel, J., Van Ness, P., Henry, T. and Crandall, P.: 1989, Ictal localization of temporal lobe seizures with scalp/spenoidal recordings, *Neurology* **39**(10), 1288–1293.

Rodin, E. and Modur, P.: 2008, Ictal intracranial infraslow EEG activity, *Clinical Neurophysiology* **119**, 2188–2200.

Rosenow, F., Hamer, H., Knake, S., Katsarou, N., Fritsch, B., Oertel, W., K.Shiratori and Lüders, H.: 2001, Lateralisierende und lokalisierende Anfallssymptome. Bedeutung und Anwendung in der klinischen Praxis, *Der Nervenarzt* **72**(10), 743–749.

Rosenow, F. and Lüders, H.: 2001, Presurgical evaluation of epilepsy, *Brain* **124**, 1683–1700.

Salinksy, M., Kanter, R. and Dasheiff, R.: 1987, Effectiveness of Multiple EEGs in Supporting the Diagnosis of Epilepsy: An Operational Curve, *Epilepsia* **28**(4), 331–334.

Sameshima, K. and Baccala, L.: 1999, Using partial directed coherence to describe neuronal ensemble interactions, *Journal of Neuroscience Methods* **94**, 93–103.

Sanei, S. and Chambers, J. (eds): 2007, *EEG Signal Processing*, Wiley.

Scharf, L. L.: 1991, *Statistical Signal Processing*, Addison-Wesley.

Scharf, L. L. and Friedlander, B.: 1994, Matched Subspace Detectors, *IEEE Transactions on Signal Processing* **42**(8), 2146–2157.

Scharf, L. L. and Lytle, D. W.: 1971, Signal Detection in Gaussian Noise of Unknown Level: An Invariance Application, *IEEE Transactions on Information Theory* **17**(4), 404–411.

Schelter, B., Timmer, J. and Eichler, M.: 2009, Assessing the strength of directed influences among neural signals using renormalized partial directed coherence, *Journal of Neuroscience Methods* **179**(1), 121–130.

Schelter, B., Winterhalder, M., Eichler, M., Peifer, M., Hellwig, B., Guschlbauer, B., Lücking, C., Dahlhaus, R. and Timmer, J.: 2005, Testing for directed influences among neural signals using partial directed coherence, *Journal of Neuroscience Methods* **152**, 210–219.

Schiller, Y. and Najjar, Y.: 2008, Quantifying the response to antiepileptic drugs, *Neurology* **70**, 54–65.

Schindler, K., Elger, C. and Lehnertz, K.: 2007, Increasing synchronization may promote seizure termination: Evidence from status epilepticus, *Clinical Neurophysiology* **118**, 1955–1968.

Schindler, K., Leung, H., Elger, C. and Lehnertz, K.: 2007, Assessing seizure dynamics by analysing the correlation structure of multichannel intracranial EEG, *Brain* **130**, 65–77.

Schlögl, A., Keinrath, C., Zimmermann, D., Scherer, R., Leeb, R. and Pfurtscheller, G.: 2007, A fully automated correction method of EOG artifacts in EEG recordings, *Clinical Neurophysiology* **118**, 98–104.

Schlögl, A. and Supp, G.: 2006, Analyzing event-related EEG data with multivariate autoregressive parameters, *in* C. Neuper and W. Klimesch (eds), *Progress in brain research*, Elsevier, pp. 135–147.

Schmidt, D. and Sillanpää, M.: 2012, Evidence-based review on the natural history of the epilepsies, *Current Opinion in Neurology* **25**, 159–163.

Schneble, H.: 2003, *Heillos, heilig, heilbar. Die Geschichte der Epilepsie von den Anfängen bis heute*, Walter de Gruyter.

Schneider, T. and Neumaier, A.: 2001, Algorithm 808: ARfit, a Matlab package for the estimation of parameters and eigenmodes of multivariate autoregressive models, *ACM Transactions on Mathematical Software* **27**(1), 58–65.

Schuele, S. and Lüders, H.: 2008, Intractable epilepsy: management and therapeutic alternatives, *Lancet Neurology* **7**, 514–524.

Schulz, R., Lüders, H., Hoppe, M., Tuxhorn, I., May, T. and Ebner, A.: 2000, Interictal EEG and Ictal Scalp EEG Propagation Are Highly Predictive of Surgical Outcome in Mesial Temporal Lobe Epilepsy, *Epilepsia* **41**(5), 564–570.

Schulze-Bonhage, A.: 2009, Deep Brain Stimulation: A New Approach to the Treatment of Epilepsy, *Deutsches Ärzteblatt International* **106**(24), 407–412.

Schuster, A.: 1898, On the Investigation of Hidden Periodicities with Application to a Supposed Twenty-Six-Day Period of Meterological Phenomena, *Terrestrial Magnetism* **3**(1), 13–41.

Schwarz, G.: 1978, Estimating the dimension of a model, *The Annals of Statistics* **6**(2), 461–464.

Seth, A. K.: 2010, A MATLAB toolbox for Granger causal connectivity analysis, *Journal of Neuroscience Methods* **186**, 262–273.

Sieghart, W. and Sperk, G.: 2002, Subunit composition, distribution and function of GABA(A) receptor subtypes, *Current Topics in Medicinal Chemistry* **2**(8), 795–816.

Smart, O., Worrell, G., Vachtsevanos, G. and Litt, B.: 2005, Automatic Detection of High Frequency Epileptiform Oscillations from Intracranial EEG Recordings of Patients with Neocortical Epilepsy, *Proceedings of the IEEE Region 5 and IEEE Denver Section Technical, Professional and Student Development Workshop* pp. 53–58.

Smith, J.: 2005, EEG in the diagnosis, classification, and management of patients with epilepsy, *Journal of Neurology, Neurosurgery & Psychiatry* **76**(Suppl II), ii2–ii7.

Sommerlade, L., Eichler, M., Jachan, M., Henschel, K., Timmer, J. and Schelter, B.: 2009, Estimating causal dependencies in networks of nonlinear stochastic dynamical systems, *Physical Review E* **80**, 051128.

Sommerlade, L., Henschel, K., Wohlmuth, J., Jachan, M., Amtage, F., Hellwig, B., Lücking, C. H., Timmer, J. and Schelter, B.: 2009, Time-variant estimation of directed influences during Parkinsonian tremor, *Journal of Physiology - Paris* **103**(6), 348–352.

Sommerlade, L., Thiel, M., Platt, B., Planod, A., Riedel, G., Grebogic, C., Timmer, J. and Schelter, B.: 2012, Inference of Granger causal time-dependent influences in noisy multivariate time series, *Journal of Neuroscience Methods* **203**, 173–185.

Spencer, S.: 1996, Long-Term Outcome After Epilepsy Surgery, *Epilepsia* **37**(9), 807–813.

Staba, R., Wilson, C., Bragin, A., Fried, I. and Engel, J.: 2002, Quantitative Analysis of High-Frequency Oscillations (80-500 Hz) Recorded in Human Epileptic Hippocampus and Entorhinal Cortex, *Journal of Neurophysiology* **88**, 1743–1752.

Stedman (ed.): 2005, *Stedman's Medical Dictionary for the Health Professions & Nursing*, Lippincott Williams & Wilkins.

Stefan, H., Hopfengärtner, R., Kreiselmeyer, G., Weigel, D., Rampp, S., Kerling, F., Blümcke, I. and Buchfelder, M.: 2008, Interictal triple ECoG characteristics of temporal lobe epilepsies: An intraoperative ECoG analysis correlated with surgical outcome, *Clinical Neurophysiology* **119**, 642–652.

Takahashi, D., Baccala, L. and Sameshima, K.: 2007, Connectivity inference between Neural Structures via Partial Directed Coherence, *Journal of Applied Statistics* **34**(10), 1259–1273.

Takahashi, D., Baccala, L. and Sameshima, K.: 2010, Information theoretic interpretation of frequency domain connectivity measures, *Biological Cybernetics* **103**, 463–469.

Tavee, J.: 2010, Neurology, *in* W. D. Carey (ed.), *Current Clinical Medicine*, Saunders Elsevier, chapter 10.

Terrien, J., Hassan, M., Marque, C. and Karlsson, B.: 2008, Use of piecewise stationary segmentation as a pre-treatment for synchronization measures, *Proceedings of the IEEE EMBC* pp. 2662–2664.

Thammongkol, S., Vears, D., Bicknell-Royle, J., Nation, J., Draffin, K., Stewart, K., Scheffer, I. and Mackay, M.: 2012, Efficacy of the ketogenic diet: Which epilepsies respond?, *Epilepsia* **53**(3), e55–e59.

The MathWorks, Inc.: 2010, *Matlab® R2010b: online help*.

Thorbecke, R. and Pfäfflin, M.: 2012, Social aspects of epilepsy and rehabilitation, *in* H. Stefan and W. Theodore (eds), *Handbook of Clinical Neurology*, Vol. 108, Elsevier, chapter 60, pp. 984–999.

Thurstone, L.: 1947, *Multiple factor analysis*, University of Chicago Press.

Tibshirani, R.: 1996, Regression Shrinkage and Selection via the Lasso, *Journal of the Royal Statistical Society B* **58**(1), 267–288.

Titchmarsh, E.: 1962, *Introduction to the theory of Fourier integrals*, Oxford University Press.

Téllez-Zenteno, J., Dhar, R. and Wiebe, S.: 2005, Long-term seizure outcomes following epilepsy surgery: a systematic review and meta-analysis, *Brain* **128**, 1188–1198.

Tong, S. and Thakor, N. V. (eds): 2009, *Quantitative EEG analysis methods and clinical applications*, Artech House.

Tseng, S.-Y., Chen, R.-C., Chong, F.-C. and Kuo, T.-S.: 1995, Evaluation of parametric methods in EEG signal analysis, *Medical Engineering & Physics* **17**, 71–78.

Urrestarazu, E., Chander, R., Dubeau, F. and Gotman, J.: 2007, Interictal high-frequency oscillations (100-500 Hz) in the intracerebral EEG of epileptic patients, *Brain* **130**, 2354–2366.

Ursulean, R. and Lazar, A.: 2007, Segmentation of Electroencephalographic Signals using an Optimal Orthogonal Linear Prediction Algorithm, *Electronics and Electrical Engineering* **1**, 73–76.

Usui, N., Terada, K., Baba, K., Matsuda, K., Nakamura, F., Usui, K., Tottori, T., Umeoka, S., Fujitani, S., Mihara, T. and Inoue, Y.: 2010, Very high frequency oscillations (over 1000 Hz) in human epilepsy, *Clinical Neurophysiology* **121**, 1825–1831.

Usui, N., Terada, K., Baba, K., Matsuda, K., Nakamura, F., Usui, K., Yamaguchi, M., Tottori, T., Umeoka, S., Fujitani, S., Kondo, A., Mihara, T. and Inoue, Y.: 2011, Clinical significance of ictal high frequency oscillations in medial temporal lobe epilepsy, *Clinical Neurophysiology* **122**, 1693–1700.

Van Gompel, J. J., Worrell, G. A., Bell, M. L., Patrick, T. A., Cascino, G. D., Raffel, C., Marsh, W. R. and Meyer, F. B.: 2008, Intracranial

Electrocencephalography with Subdural Grid Electrodes: Techniques, Complications, and Outcomes, *Neurosurgery* **63**, 498–506.

van Mierlo, P., Carrette, E., Hallez, H., Vonck, K., Van Roost, D., Boon, P. and Staelus, S.: 2011, Epileptogenic focus localization through connectivity analysis of the intracranial EEG: A retrospective study in 2 patients, *Proceedings of the IEEE EMBS Conference on Neural Engineering* pp. 655–658.

van Putten, M., Kind, T., Visser, F. and Lagerburg, V.: 2005, Detecting temporal lobe seizures from scalp EEG recordings: A comparison of various features, *Clinical Neurophysiology* **116**, 2480–2489.

van 't Klooster, M., Zijlmans, M., Leijten, F., Ferrier, C., van Putten, M. and Huiskamp, G.: 2011, Time-frequency analysis of single pulse electrical stimulation to assist delineation of epileptogenic cortex, *Brain* **134**, 2855–2866.

Vanhatalo, S., Voipio, J. and Kaila, K.: 2005, Full-band EEG (FbEEG): an emerging standard in electroencephalography, *Clinical Neurophysiology* **116**, 1–8.

Varsavsky, A. and Mareels, I.: 2006, Patient Un-Specific Detection of Epileptic Seizures Through Changes in Variance, *Proceedings of the IEEE EMBC* pp. 3747–3750.

Varsavsky, A. and Mareels, I.: 2007, A Complete Strategy for Patient Un-specific Detection of Epileptic Seizures Using Crude Estimations of Entropy, *Proceedings of the IEEE EMBC* pp. 6491–6494.

Varsavsky, A., Mareels, I. and Cook, M.: 2011, *Epileptic Seizures and the EEG. Measurement, Models, Detection and Prediction*, CRC Press.

Vigario, R. N.: 1997, Extraction of ocular artefacts from EEG using independent component analysis, *Electroencephalography and Clinical Neurophysiology* **103**, 395–404.

Ville, J.: 1948, Théorie et application de la notion de signal analytique, *Cables et transmission* **2**(1), 61–74.

von Ellenrieder, N., Andrade-Valenca, L., Dubeau, F. and Gotman, J.: 2012, Automatic detection of fast oscillations (40-200 Hz) in scalp EEG recordings, *Clinical Neurophysiology* **123**, 670–680.

Vonck, K., de Herdt, V., Sprengers, M. and Ben-Menachem, E.: 2012, Neurostimulation for epilepsy, *in* H. Stefan and W. Theodore (eds), *Handbook of Clinical Neurology*, Vol. 108, Elsevier, chapter 58, pp. 956–970.

Walczak, T., Radtke, R. and Lewis, D.: 1992, Accuracy and interobserver reliability of scalp ictal EEG, *Neurology* **42**(12), 2279–2285.

Wallstrom, G., Kass, R., Miller, A., Cohn, J. and Fox, N.: 2004, Automatic correction of ocular artifacts in the EEG: a comparison of regression-based and component-based methods, *International Journal of Psychophysiology* **53**, 105–119.

Warren, C., Hu, S., Stead, M., Brinkmann, B., Bower, M. and Worrell, G.: 2010, Synchrony in Normal and Focal Epileptic Brain: The Seizure Onset Zone is Functionally Disconnected, *Journal of Neurophysiology* **104**, 3530–3539.

Weinand, M., Wyler, A., Richey, E., Phillips, B. and Somes, G.: 1992, Long-term ictal monitoring with subdural strip electrodes: prognostic factors for selecting temporal lobectomy candidates, *Journal of Neurosurgery* **77**, 20–28.

Welch, P.: 1967, The Use of Fast Fourier Transform for the Estimation of Power Spectra. A Method Based on Time Averaging Over Short, Modified Periodograms, *IEEE Transactions on Audio and Audioacoustics* **Au-15**(2), 70–73.

Wendling, F., Bartolomei, F., Bellanger, J. and Chauvel, P.: 2002, Epileptic fast activity can be explained by a model of impaired GABAergic dentritic inhibition, *European Journal of Neuroscience* **15**, 1499–1508.

Wiebe, S., Blume, W., Girvin, J. and Eliasziw, M.: 2001, A randomized, controlled trial of surgery for temporal-lobe epilepsy, *New England Journal of Medicine* **345**(5), 311–318.

Wiener, N.: 1956, The theory of prediction, Vol. 1, chapter 8.

Wieser, H. and Yasargil, M.: 1982, Selective amygdalahippocampectomy as a surgical treatment of mesiobasal limbic epilepsy, *Surgical Neurology* **17**, 445–457.

Wiesner, H., Blume, W., Fish, D., Goldensohn, E., Hufnagel, A., King, D., Sperling, M. and Lüders, H.: 2001, Proposal for a New Classification of Outcome with Respect to Epileptic Seizures Following Epilepsy Surgery, *Epilepsia* **42**(2), 282–286.

Wilke, C., Ding, L. and He, B.: 2008, Estimation of Time-Varying Connectivity Patterns Through the Use of an Adaptive Directed Transfer Function, *IEEE Transactions on Biomedical Engineering* **55**(11), 2557–2564.

Willoughby, J., Fitzgibbon, S., Pope, K., Mackenzie, L., Medvedev, A., Clark, C., Davey, M. and Wilcox, R.: 2003, Persistent abnormality detected in the non-ictal electrocencephalogram in primary generalised epilepsy, *Journal of Neurology, Neurosurgery & Psychiatry* **74**, 51–55.

Winterhalder, M., Schelter, B., Hesse, W., Schwab, K., Leistritz, L., Klan, D., Bauer, R., Timmer, J. and Witte, H.: 2005, Comparison of linear signal processing techniques to infer directed interactions in multivariate neural systems, *Signal Processing* **85**, 2137–2160.

Worrell, G., Gardner, A., Stead, S., Hu, S., Goerss, S., Cascino, G., Meyer, F., Marsh, R. and Litt, B.: 2008, High-frequency oscillations in human temporal lobe: simultaneous microwire and clinical macroelectrode recordings, *Brain* **131**, 928–937.

Worrell, G., Jerbi, K., Kobayashi, K., Lina, K., Zelmann, R. and Le Van Quyen, M.: 2012, Recording and analysis techniques for high-frequency oscillations, *Progress in Neurobiology* **98**, 265–278.

Wu, J., Koh, S., Sankar, R. and Mathern, G.: 2008, Paroxysmal fast activity: An interictal scalp EEG marker of epileptogenesis in children, *Epilepsy Research* **82**, 99–106.

Wu, J., Sankar, R., Lerner, J., Matsumoto, J., Vinters, H. and Mathern, G.: 2010, Removing interictal fast ripples on electrocorticography linked with seizure freedom in children, *Neurology* **75**, 1686–1694.

Wu, L. and Gotman, J.: 1998, Segmentation and classification of EEG during epileptic seizures, *Electroencephalography and Clinical Neurophysiology* **106**, 344–356.

Yule, G. U.: 1927, On a Method of Investigating Periodicities in Disturbed Series, with Special Reference to Wolfer's Sunspot Numbers, *Philosophical Transactions of the Royal Society of London. Series A, Containing Papers of a Mathematical or Physical Character* **226**, 267–298.

Zelmann, R., Mari, F., Jacobs, J., Zijlmans, M., Chander, R. and Gotman, J.: 2010, Automatic detector of High Frequency Oscillations for human recordings with macroelectrodes, *Proceedings of the IEEE EMBC* pp. 2329–2333.

Zelmann, R., Mari, F., Jacobs, J., Zijlmans, M., Dubeau, F. and Gotman, J.: 2012, A comparison between detectors of high frequency oscillations, *Clinical Neurophysiology* **123**, 106–116.

Zelmann, R., Zijlmans, M., Jacobs, J., Chatillon, C.-E. and Gotman, J.: 2009, Improving the identification of High Frequency Oscillations, *Clinical Neurophysiology* **120**, 1457–1464.

Zijlmans, M., Jacobs, J., Kahn, Y., Zelmann, R. and Dubeau, F.: 2011, Ictal and interictal high frequency oscillations in patients with focal epilepsy, *Clinical Neurophysiology* **122**, 664–671.

Zijlmans, M., Jacobs, J., Zelmann, R., Dubeau, F. and Gotman, J.: 2009a, High frequency oscillations and seizure frequency in patients with focal epilepsy, *Epilepsy Research* **85**, 287–292.

Zijlmans, M., Jacobs, J., Zelmann, R., Dubeau, F. and Gotman, J.: 2009b, High-frequency oscillations mirror disease activity in patients with epilepsy, *Neurology* **72**, 979–986.

Zoldi, S., Krystal, A. and Greenside, H.: 2000, Stationarity and Redundancy of Multichannel EEG Data Recorded During Generalized Tonic-Clonic Seizures, *Brain Topography* **12**(3), 187–200.

Zumsteg, D. and Wieser, H. G.: 2000, Presurgical Evaluation: Current Role of Invasive EEG, *Epilepsia* **41**, S55–S60.

# Index

## Symbols

10-10-system . . . . . . . . . . . . . . . . . . 34
10-20-system . . . . . . . . . . . . . . . . . . 34

## A

absence . . . . . . . . . . . . . . . . . . . . . . . 47
action potential . . . . . . . . . . . . . . . . 30
adaptive estimation . . . . . . . . . . . . 67
admissible . . . . . . . . . . 167, 174, 176
adverse event . . . . . . . . . 51, 61, 229
AED . . . . . . . . . . . . . . 22, 51, 56, 61
AIC . . . . . . . . . . . . . . . . . . . . 144, 152
amygdala-hippocampectomy . . . 58
An algorithm . . . . . . . . . . . . . . . . 144
anamnesis . . . . . . . . . . . 49, 117, 212
       seizure . . . . . . . . . . . . . . . . . . . . . 49
       social . . . . . . . . . . . . . . . . . . . . . . 49
anti-epileptic drug . . . . . . . *see* AED
aphasia . . . . . . . . . . . . . . . . . . . . . . 213
AR model 75, 77 ff., 107, 114 f., 129, 139 ff., 143, 149, 151, 153, 166 f., 171, 176, 248
       coefficient . . . 76 f., 86, 153, 246
       order fitting . . . . . . . . . . 164, 184
       regular . . . . . . . . . . . 75, 106, 166
arrow 117, 154, 162, 165, 175, 177, 182

map . . . . . . . . . . . . . . . . . 154, 177
artifact . . . . . . . . . . 40, 189, 192, 224
       blink . . . . . . . . . . . . . . . . . 40, 224
       ECG . . . . . . . . . . . . . . . . . . . . . . 41
       electrode or cable . . . . . . . . . . 42
       electrostatic induction . . . . . . 42
       eye movement . . . . . . . . . . . . . 40
       galvanic skin . . . . . . . . . . . . . . 41
       glossocinetic . . . . . . . . . . . . . . 41
       line interference . . . . . . . . . . . 42
       muscle . . . . . . . . . . 40, 224, 227
       physiological . . . . . . . . . . . . . . 40
       pulsation . . . . . . . . . . . . . . . . . 41
       removal . . . . . . . . . . . . . . . . . 224
       technical . . . . . . . . . . . . . . . . . 42
aura . . . . . . . . . . . . . . . . . . . . . . . . . 47
autoregressive model *see* AR model

## B

Band Power Measure . . . . *see* BPM
bias 66, 70, 72 f., 142, 236, 239, 244
       asymptotic 70, 72, 74, 237, 239, 245
BIC . . . . . . . . . . . . . . . . . . . . 144, 172
bootstrapping . . . . . . . . . . . . . 84, 149
BPM . . . . . . . . . . . . . . 189, 196, 200
brain . . . . . . . . . . . . . . . . . . . . . . . . 28

## C

cavernoma . . . . . . . . . . . . . . . . . . . 36
cerebrum . . . . . . . . . . . . . . . . . . . . . 43
cetogenic diet . . . . . . . . . . . . . . . . . 52
channel selection . . . . . . . . 117, 140, 143, 148, 152, 161, 164, 167, 169 f., 174, 180
channel set
    extrinsic . . . . . . 141 f., 144, 161 f.
classification . 117, 185, 192, 198 f., 205 f.
clinical signs . *see* lateralizing signs
co-movement . 105, 165 f., 168, 180, 182
comorbidity . . . . . . . . . . . . . . . . . . 61
concordance principle . . . . . . . . . . 20
confidence interval . . . . . . . . . 81, 83
    frequency-dependent . . . 89, 151
consciousness . . . . . . . . . . 44, 47, 117
consistency . . . . . . . 66, 72, 142, 240
contribution
    frequency . . . . 127 f., 188 f., 203
    intrinsic . . . . . . . . . 142, 146, 161
    partial . . . . . . . . . . 141, 145, 147
    partial extrinsic . . 142, 158, 161
    total extrinsic . . . . . . . . . . . . . 142
convolution . . . . . . . . . . . . . . 68, 245
correlation estimator . . . 67, 70, 238, 243
correlogram . . . 70, 236 f., 239, 242
    Blackman-Tukey . . . . . . . . . . . 71
cortex . . . . . . . . . . . 28, 43 f., 53, 163
    insular . . . . . . . . . . . . . . . . . . . 29
    visual . . . . . . . . . . . . . . . . . . . 223
cortical zone . . . . . . . . . . . . . . . . . 52
    epileptogenic zone . . . 53, 55, 58
    functional deficit zone . . 53, 213
    irritative zone . . . . . . . . . . . 52, 54
    seizure onset zone . . . . *see* SOZ
    symptomatogenic zone . . 52, 57
coupled frequency . . . . . . . . . . 191 f.
coupling . . . *see* dependeny measure
coverage . . . . . . . . . . . . . . . . 219, 228
craniotomy . . . . . . . . . . . . . . . 43, 229
cranium . . . . . . 28, 43, 55, 225, 229
CT . . . . . . . . . . . . . . . . . . . . . . . . . . 50

## D

decision rule . . . . . . . . . 97, 102, 104
dependency measure 78, 115, 139 f., 151, 153, 158, 180, 228
    direct . . . . . . . . . . . . . . . . . . . . 79
    directed 79, 145 f., 154, 162, 165
    indirect . . . . . . . . . . . . . . . . . . . 79
    of linear dependence . . . . . . . . 91
    undirected . . . . . . . . . . . . . . . . 79
depression . . . . . . . . . . . . . . . . . . . 61
diagnosis . . . . . . . . . . . . . . . . . . . . 49
    differential . . . . . . . . . . . . . . . . 49
diagnostic imaging 50, 53, 117, 214, 221
dichotomy . . . . . . . . . . . . . . . 99, 253
Dirac delta . . . . . 134, 237, 240, 246
directed coherence . . . . . . . . . . . . 84

directed transfer function . *see* DTF
directedness . . . . . . . 79, 84, 87 f., 91
Dirichlet kernel . . . 68, 71, 235, 237, 239
discrete-time Fourier transform . *see* DTFT
distinctive constellation . . . . . . . . 48
downsampling . . . . . . . . . . . 119, 126
DTF . . . . . . . . . . 83, 89, 91, 140, 248
   direct (dDTF) . . . . . . . . . . . . . . 85
   full frequency (ffDTF) . . . . . . 85
   short-term (sDTF) . . . . . . . . . . 85
DTFT . . . . . . . . . . . . . 64, 68, 76, 235

# E

ECoG . . . 43, 54, 60, 112, 119, 125, 130, 139 f., 142, 146, 153, 161, 166, 180, 184, 188, 190, 195 f., 198, 200, 205, 214
   intraoperative . . . . . . . . . . . . . . 58
edge . . . . . . . . . . . . . . . . . . . . 79, 139
EEG . . . . . . . . . . 22, 33, 38 f., 50, 52, 54, 57, 65, 79, 105, 112, 117, 166, 186, 190, 213, 223, 228
   diagnostic value . . . . . . . . . . . . 50
   high-density . . . . . . . . . . . . . . 229
   intracranial . . . . . . . . . . . 43, 112
   invasive 43, 52, 54, 58, 224, 229
eigenvalue . . . . . . . . . . . . . . . . . . 106
EIPR . 141, 146 ff., 151, 153 f., 158, 160, 162
electrocorticography . . . . *see* ECoG

electrode
   depth 43, 57, 112, 125, 220, 229
   macro . . . . . . . . . . . . . . . . . . . . 55
   micro . . . . . . . . . . . . . . . . . . . . 55
   spenoidal . . . . . . . . . . . . . . . . . 43
   subdural . . . . . 57, 118, 125, 219
      grid . . . . . . . . . . . . . . . . 43, 229
      strip . . . . . . . 43, 163, 183, 228
electroencephalogram . . . . *see* EEG
electroencephalography . . *see* EEG
EMU . . . . . . . . . . . . . . . . . . 56, 118
epilepsy . . . . . . . . . . . . . . . . . . . . 44
   course of disease . . . . . . . . . . . 45
      relapse . . . . . . . . . . . . . . . . . . 46
      remission . . . . . . . . . . . . . . . 46
   focal 49, 117, 162, 181, 206, 213, 216
   incidence . . . . . . . . . . . . . . . . . 45
   prevalence . . . . . . . . . . . . . . . . 45
   surgery . . . . . . . . . . . *see* surgery
   therapy-resistant . . . . . . . 56, 117
Epilepsy Monitoring Unit *see* EMU
etiology . . . . . . . . . . . . . . . . . . 47, 49
   genetic . . . . . . . . . . . . . . . . . . . 48
   metabolic . . . . . . . . . . . . . . . . . 48
   structural . . . . . . . . . . . . . . 48, 50
extrinsic-to-intrinsic-power-ratio
   *see* EIPR

# F

factor . . . . . . . . . . . . . . . . . . . . . 105
   static . . . . . . . . . . 106, 168, 171 f.

factor loading matrix 106, 168, 172, 181
factor model . . . . . . . 105, 117, 165 f.
    quasi-static . . . . . . . . . . . . . . . 106
feedback process . 31, 163, 182, 228
Fejer kernel . . . . . . . . . . . . . 237, 239
field potential . . . . . . . . . . . . . . . . 38
filter
    band-pass . . . . . . . 124, 128, 134
    coefficient . . . . . . . . . . . . . . . . 134
    high-pass . . . . . . . . 119, 128, 135
    Kalman . . . . . . . . . . . . . . . . . . 114
    low-pass . . . . . . . . . 55, 119, 126
    matched subspace . *see* matched subspace filter
    notch . . . . . . . . . . . . 42, 119, 126
fissure . . . . . . . . . . . . . . . . . . . . . 28 f.
    calcarine . . . . . . . . . . . . . . . . . . 29
    central . . . . . . . . . . . . . . . . . . . . 29
    cingulate . . . . . . . . . . . . . . . . . . 29
    inter-hemispheric . . . . . . . . . . . 28
    lateral . . . . . . . . . . . . . . . . . . . . 29
    parietooccipital . . . . . . . . . . . . 29
fMRI . . . . . . . . . . . . . . . . . . . 50, 212
Fourier transform . . . . . . . *see* DTFT
frequency band . . 33, 124, 135, 140, 153, 176, 189, 196, 228

## G

GABA . . . . . . . . . . . . . . . . . . . 30, 51
generalized function . . . . . . . . . . 237
glutamate . . . . . . . . . . . . . . . . . 30, 51

gold standard . . . . . . . . . . . . . . . 117
Granger causality . . 76, 84, 92, 158, 165, 169, 171, 181, 184, 228, 248
    conditional . 77, 85, 89, 91, 148, 166, 168, 246, 250
    frequency domain . . . 78, 87, 90, 146
    index . . . . . . . . . . . . . . . . . . . . . 91
graph . . . . . . . . . . . . . . 79, 139, 165
gyrus . . . . . . . . . . . . . . . . . . . . . . . 28

## H

head . . . . . . . . . . . . . . . . . . . . . . . . 27
health . . . . . . . . . . . . . . . . . . . . . . 61
hemisphere . . . . . 28, 57, 220 f., 227
    language . . . . . . . . . . . . . . . . . 213
HFO . . . 54, 60, 115, 119, 123, 196, 214, 225, 228
    detection . . . . . . . 123, 126, 135 f.
    ultra-high . . . . . . . . . . . . . . . . . 55
high frequency oscillations *see* HFO
Hilbert transform . . . . . . . . 125, 129
hippocampus . . . . . 57, 60, 163, 183
    hippocampal sclerosis . . . 49, 58
Hippocrates . . . . . . . . . . . . . . . . . 19
history . . . . . . . . . . . . . *see* anmnesis
    family . . . . . . . . . . . . . . . . . . . . 49
Hughlings-Jackson . . . . . . . . . . . 22
humors . . . . . . . . . . . . . . . . . . . . . 20
hyper-cone . . . . . . . . . . . . . . . . . 104
hyper-cylinder . . . . . . . . . . . . . . 102

hyper-plane ................. 127
hypothesis .................... 93
    composite ........ 94, 96, 100
    simple .................. 94 f.
    test ................. *see* test

# I

ictal .......................... 44
ictal slow shift ............... 56
IED ...................... 52, 60
IEEE EMBS ................. 17
ILAE ........................ 22
induced transformation ........ 98
infection ..................... 48
influence analysis . 165, 169, 175 ff., 180, 214, 228
information criterion
    Akaike ............... *see* AIC
    BIC .................. *see* BIC
interictal .................... 44
interictal epileptic discharge *see* IED
invariance ............... 98, 100
    decision rule .......... 99, 251
    hypothesis testing 99 f., 102, 104, 252
    maximal invariant statistic . 99 f., 102, 104, 251, 253
    transformation group ... 98, 252
ion channel .................. 30

# J

Jasper ....................... 23

# K

Karlin-Rubin theorem 97, 100, 102, 104

# L

latent variable .. 105, 107, 166, 168, 170, 180
lateralization ...... 36, 54, 57, 213
    language ................ 212
lateralizing signs ...... 52, 57, 213
    figure of four .......... 57, 213
    version of the head ........ 213
leakage effect ......... 68, 71 f., 74
lesion .......... 48, 53, 57, 60, 214
likelihood ratio ....... 97, 102, 104
Liverpool Adverse Event Profile 61
lobe ......................... 28
    frontal .................... 29
    limbic .................... 29
    occipital .................. 29
    parietal ................... 29
    side .................. 68, 72
    temporal ................. 29
lobectomy ................ 58, 60
localization ... 36, 53, 57, 113, 139, 153, 220, 225, 229
    3D source .......... 113 f., 229
    accuracy .................. 36
    rules ..................... 38
localizing value 52, 54, 57, 115, 214
LOOM ..................... 149

## M

matched subspace filter .. 100, 102, 114, 123, 126, 134, 252 f.
  CFAR...........104, 126, 128
meninges..................28, 43
Micromed SystemPlus Evolution 118
modal matrix ............ 100, 127
Montreal method .............. 23
MRI . 48, 50, 53, 60, 153, 176, 200, 212, 220, 231
  high-resolution ......... 50, 57
  negative......... 117, 213, 216
multi-scale modeling ......... 230

## N

NDDI-E..................... 61
neuro stimulation............. 52
neuron ...................... 30
neuropsychological testing .. 49, 53, 56, 117, 212, 231
neurotransmitter............... 30
Neyman-Pearson
  criterion................... 96
  theorem................96, 98
node ........................ 79
non-linear ........... 80, 114, 184
normal equation .. *see* Yule-Walker
normalization .. 83, 87, 89, 91, 135, 146 ff., 158, 161
nuisance parameter........98, 101

## O

observation space..............93
  acceptance region ..........93
  critical region ..............93
orbit ............... 100, 102, 104
ordinary coherence ............ 80
out-degree....155, 162 f., 177, 182

## P

Paracelsus .................... 20
parameter space ....... 93, 96, 161
paresis ...................... 213
partial coherence .. 81, 85, 247, 249
partial directed coherence . *see* PDC
partialized spectrum .. 80, 247, 249
pattern at onset ................53
PCA...........106, 168, 172, 181
  sparse ................... 181
PDC 87 ff., 91, 140, 146 f., 151, 158, 161, 250
  generalized (gPDC) ........ 90
  information theoretic form .. 90
  partial coherence function...87, 249
  partial directed coherence factor (PDCF) ................ 87
Penfield ..................... 23
periodic waveform
  analysis ............. *see* PWA
  index ................ *see* PWI
periodogram ......... 71, 237, 239
  Bartlett................... 72

Daniell . . . . . . . . . . . . . . . . . . . . 72
Welch . . . . . . . . . . . . 72, 189, 241
PESOS . . . . . . . . . . . . . . . . . . . . . . 61
PET . . . . . . . . . . . . . . . . . . . . 50, 212
phase reversal . . . . . . . . . . . . . . . . 38
polarity convention . . . . . . . . . . . . 38
post-surgical outcome 57, 59 f., 112, 206, 217, 230
postictal . . . . . . . . . . . . . . . . . . . . . 44
postsynaptic
    cleft . . . . . . . . . . . . . . . . . . . . . . 30
    potential . . . . . . . . . . . . . . . . . . 30
potential difference . . . . . . . . . . . . 33
power . . . . . . . . . . . . . . . . . . . . . . . 97
    intrinsic . . . . . . . . . . . . . . . . . . 145
    partial extrinsic . . . . . . . 145, 158
    total extrinsic . . . . . . . . . . . . . 145
power spectral density . . . . *see* PSD
pre-emphasis . . . . . . . 125, 128, 135
pre-whitening . . . . . . . . . . . 125, 135
preictal . . . . . . . . . . . . . . . . . . . . . . 44
preprocessing . . . . . . . . . . . 119, 126
presurgical evaluation . . . 43, 52, 57, 112, 117, 224
    invasive phase . . . . . . . . . . . . . . 57
    non-invasive phase . . . . . . . . . . 57
Principal Component Analysis . *see* PCA
probability
    acceptance . . . . . . . . . . . . . . . . . 94
    density function . . . . . 93, 96, 98
    detection . . . . . . . . . . . *see* power

false alarm . . . . . . . . . . . . *see* size
miss . . . . . . . . . . . . . . . . . . . . . 94
propagation . . 54, 57, 113, 117, 132, 140, 188, 195, 197, 206, 214, 217, 225
    contralateral . . . . . . . . . . . 60, 112
    framework . . 111, 115, 211, 216, 223, 230
    interhemispheric . . . . . . . . . . . 60
    ipsilateral . . . . . . . . . . . . . . . . 112
    speed . . . . . . . . . . . . 60, 112, 231
PSD . . . 68, 70, 76, 147 f., 188, 191, 237, 239 f., 245
pseudo-ensemble averaging . . . . 72
pulse generator . . . . . . . . . . . . . . . 31
PWA . . . . . . . . . . . . . . . . . . . . . . . 190
PWI . . . . . . . . . . . . . . . . . . . . . 191 f.

## Q

QOLIE-10 . . . . . . . . . . . . . . . . . . . 61
quality of life . . . . . . . . . . . . . 61, 221
questionnaire . . . . . . . . . . . . . . . . . 61

## R

Receiver Operator Characteristics *see* ROC
regression . . . . . . . . . . 141, 152, 164
    penalized . . . . . . . . . . . . . . . . 164
resection . . . . . . . . . . . . . . 56, 58, 223
residual sum of squares . . . . . . . 144
rhythmic activity . . . . . . . . . . . . . 31, 53 f., 57, 119, 138, 153, 176,

184 f., 188, 190 ff., 196, 198, 203, 205, 214
ripples .......... 54, 124, 128, 130
    fast .................... 54, 126
ROC ........................ 95

## S

sample spectrum .. *see* periodogram
sampling frequency . 64, 118 f., 126, 130, 195, 220
Scree plot ...... 168, 172, 175, 184
segmentation ... 117, 185, 188, 190, 195 f., 199 f., 206, 214, 228
    segment classification ...... *see* classification
seizure ............. 117, 213, 220
    detection ................. 113
    epileptic ............... 44, 46
    febrile ................. 45, 48
    focal .............. 47, 52, 114
    generalized ................ 47
    prediction ................ 113
    propagation .... *see* propagation
    psychogenic ............... 49
    symptomatic ............... 44
    termination ............... 115
semiology .................... 46
sensitivity ......... 50, 55, 95, 207
setup ........................ 34
    bipolar ............ 35, 39, 227
    common average ........ 36, 39
    reference .............. 34, 224

sharp-wave ............. *see* spike
shrinkage ............... 141, 143
signal detection ..... 92, 123, 125 f.
signal model ... 129, 148, 151, 158, 161, 171 f., 195
signal subspace .... 100, 103 f., 128
signal-to-noise ratio ...... *see* SNR
size ...................... 94, 97
skin ......................... 27
skull ................ *see* cranium
SNR .................... 130, 161
social well-being .............. 61
SOZ 53 f., 57, 112 f., 115, 137, 139, 153, 155, 162, 181, 188, 200, 206, 214, 216, 219, 227, 229
spatial resolution 34, 42 f., 112, 225
specificity 50, 55, 95, 126, 134, 207
SPECT ................... 50, 221
spectral estimation
    non-parametric 65, 72, 114, 189, 229
    direct ............... 71, 237
    indirect .............. 70, 236
    parametric ......... 65, 76, 140
spectral resolution ..... 65, 74, 127
spectral roll-off ..... 125, 128, 135
spectrogram ................. 132
spike ....... 36, 50, 52, 54 f., 213 f.
stability condition ......... 75, 107
static factor ............. 175, 184
stationarity . 66, 115, 129, 142, 166, 184 f., 193

short-term . . . . . . . . . . . . . 66, 115
status epilepticus . . . . . . . . . . . . . . 45
stroke . . . . . . . . . . . . . . . . . . . . . . . 48
subarachnoidal space . . . . . . . . . . 28
subdural strip . . . . . . . . *see* electrode
surgery . . . . . . 52, 56, 112, 206, 221
    target-oriented . . . . . . . . . . . . . 58
surrogate data . . . . . . . . . . . . 83, 149
    leave-one-out method *see* LOOM
symmetry . . . . . . . . . . . . . . . . . . . . 98
symptom . . . . . . . . . . . . . . . . . . . . 46
    cognitive . . . . . . . . . . . . . . . . . 49
    psychiatric . . . . . . . . . . . . . . . . 49
symptomatology . . 46, 57, 117, 213
synchronization . . 32, 44, 114, 145,
    162, 181
    hyper . . . 44, 117, 119, 162, 228
syncope . . . . . . . . . . . . . . . . . . . . . 49
syndrome . . . . . . . . . . 46, 49, 59, 231
    Lennox-Gastaut . . . . . . . . . . . . 48
    West . . . . . . . . . . . . . . . . . . . . . 48

# T

tapering function . . . . . . *see* window
temperaments . . . . . . . . . . . . . . . . 20
test
    binary . . . . . . . . . . . . . . . . . . . 93
    F . . . . . . . . . . . . . . 169, 181, 203
    function . . . . . . . . . . . . . . . . . . 93
    most powerful invariant . . 98 ff.,
        102
    multiple . . . . . . . . . . . . . . . . . . 93

specification . . . . . . . . . . . . . . . 95
t . . . . . . . . . . . . . . . . . . . . . . . . 150
uniformly most powerful . . . . 97
therapy . . . . . . . . . . . . . . . . . . . 22, 51
    combination . . . . . . . . . . 51, 118
    mono . . . . . . . . . . . . . . . . . . . . 51
    poly . . . . . . . . . . . . . . . . . . . . . 51
threshold . . . . . . . . . . . . 30, 95, 125,
    129 f., 132, 149 ff., 158, 162,
    169, 171, 174, 176, 186, 189,
    193, 196, 198, 204
    exceeding . 131 f., 135, 196, 203
Tissot . . . . . . . . . . . . . . . . . . . . . . . 21
TLE . . . . . 34, 49 f., 53, 60, 112, 190
transfer function . . . . . . . . . . 75, 128
trauma . . . . . . . . . . . . . . . . . . . . . . 48
tumor . . . . . . . . . . . . . . . . . . . . 48, 60

# V

vertex . . . . . . . . . . . . . . . 79, 139, 165
video-EEG monitoring . . 36, 50, 56,
    118, 212, 219, 224
visual inspection . 57, 113, 117, 121,
    125, 133, 155, 177, 182, 192,
    199, 205 f., 214, 216, 224,
    227 f.
von Leyden . . . . . . . . . . . . . . . . . . 21

# W

well-definedness . . . . . . . . . . . . . 250
West . . . . . . . . . . . . . . . . . . . . . . . . 21
white noise . 75, 101, 107, 130, 149,
    172, 191, 195, 240

window 68, 115, 142, 153, 175, 184, 236, 244
    Bartlett . . . . . . . . . . . 69, 236, 239
    Hamming . . . . . . . . . . . . . . 69, 74
    rectangular . . 67, 71 f., 235, 239
    sliding . . . . . . . . . . . . . . . 186, 189

World Health Organization . . . . . 61

## Y

Yule-Walker equation . . . . . . 76, 142

## Z

z-transform . . . . . . . . . . . . . . . . . . 76

# i want morebooks!

Buy your books fast and straightforward online - at one of world's fastest growing online book stores! Environmentally sound due to Print-on-Demand technologies.

Buy your books online at

## www.get-morebooks.com

Kaufen Sie Ihre Bücher schnell und unkompliziert online – auf einer der am schnellsten wachsenden Buchhandelsplattformen weltweit! Dank Print-On-Demand umwelt- und ressourcenschonend produziert.

Bücher schneller online kaufen

## www.morebooks.de

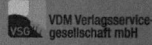 VDM Verlagsservicegesellschaft mbH
Heinrich-Böcking-Str. 6-8   Telefon: +49 681 3720 174   info@vdm-vsg.de
D - 66121 Saarbrücken   Telefax: +49 681 3720 1749   www.vdm-vsg.de

Printed by Books on Demand GmbH, Norderstedt / Germany